Sun Tzu and Information Warfare

Sun Tzu amd Information Warfare

Text in Bembo Semibold
Titles in Bernhard Bold Condensed
Book design by Mary A. Sommerville
Cover design by Juan Medrano and Elizabeth Woodard

The Sun Tzu Art of War in Information Warfare research competition is named in honor of Sun Tzu, one of the earliest great military thinkers who realized that war, a matter of vital importance to the State, demanded study and analysis. His works (circa late 4th century B.C.) are the first known attempt to formulate a rational basis for the planning and conduct of military operations. His purpose, according to Samuel B. Griffith, was "to develop a systematic treatise to guide rulers and generals in the intellegent prosecution of successful war." Sun Tzu was also convinced that careful planning based on sound *information* would contribute to speedy victory. It is from these intellectual and historical roots that the Sun Tzu Art of War in Information Warfare research competition draws its *raison d'être*.

Dr. Robert E. Neilson, Editor
Information Resources Management College
National Defense University
Ft. McNair, Washington, DC

Funding for the Sun Tzu Art of War in Information Warfare research competition is provided by the National Defense University Foundation, Fort Lesley J. McNair, Washington, DC

Sun Tzu and Information Warfare

A collection of winning papers from the Sun Tzu Art of War in Information Warfare Competition

Edited by
Robert E. Neilson
Information Resources Management College

1997
National Defense University Press
Washington, DC

National Defense University Press Publications

To increase general knowledge and inform discussion, the Institute for National Strategic Studies, through its publication arm the NDU Press, publishes *Strategic Forums*; McNair Papers; proceedings of University- and Institute-sponsored symposia; books relating to U.S. national security, especially to issues of joint, combined, or coalition warfare, peacekeeping operations, and national strategy; and a variety of other works designed to circulate contemporary comment and offer alternatives to current policy. The Press occasionally publishes out-of-print defense classics, historical works, and other especially timely or distinguished writing on national security.

CONTENTS

PREFACE by the Editor   *ix*

KNOWLEDGE STRATEGIES: Balancing Ends, Ways, and Means in the Information Age
  William R. Fast   *3*

THE SILICON SPEAR: An Assessment of Information Based Warfare and U.S. National Security
  Charles B. Everett, Moss Dewindt, & Shane McDade   *33*

INFORMATION TERRORISM: Can You Trust Your Toaster?
  Matthew G. Devost, Brian K. Houghton,
  & Neal A. Pollard   *63*

INFORMATION WARFARE: The Organizational Dimension
  Brian Fredericks   *79*

A CHAPTER NOT YET WRITTEN: Information Management and the Challenge of Battle Command
  Adolph Carlson   *103*

UNINTENDED CONSEQUENCES OF JOINT DIGITIZATION
  Steven G. Fox   *125*

INFORMATION WARFARE: Issues and Perspectives
  John H. Miller   *145*

PREFACE

Sun Tzu and Information Warfare contains papers submitted by authors in response to an open international research competition sponsored by the Information Resources Management College, National Defense University and funded by the National Defense University Foundation. Papers published in this collection include winners of the 1995 and 1996 competitions. The purpose of the competition is to stimulate innovative thought on the oft debated subject of information warfare (IW). (See the back pages of this volume or http://www.ndu.edu for further information regarding the purpose, eligibility, and evaluation criteria of the Sun Tzu Award.)

As a discrete subject, information warfare has received increasing attention from politicians, scientists, academics, futurists, military strategists, warfighters, logisticians, and the media. Much of this increased attention revolves around salient issues including:

- definition of information warfare—establishing conceptual and operational boundaries.
- legal environment for information age conflict
- doctrinal issues and force structure implications
- organizational implications—DOD, Federal and private sector
- new environments for security affairs and conflict
- relationship with the Revolution in Military Affairs (RMA)
- changing political and social milieus
- National Information Infrastructure (NII) and infrastructure security—implications for strategic vulnerability
- national policy guidance—the vacuum
- the defense planning process in the information age
- "mapping" cyberspace
- the impact of the information age on the intelligence community

- historical evolution of IW
- non-linearity, complexity, and chaos theory—conceptual links to the information age.

The papers contained in this collection address several of the issue areas mentioned above, providing innovative and provocative thought to foster a continuing dialogue between interested parties who have interest in information warfare as an integral part of national security strategy.   The 1996 winners of the Sun Tzu Award include "Knowledge Strategies: Balancing Ends, Ways, and Means in the Information Age," by LTC William Fast, which describes the effects of information age technologies on United States values, national interests, security policy, and how the *ends, ways, and means* paradigm must adapt to information age warfare. Matthew Devost, Brian Houghton, and Neil Pollard, in "Information Terrorism: Can You Trust Your Toaster?" present a futuristic information warfare scenario and an information terror typology which illustrates the lethality of information terrorism attacks.   "The Silicon Spear: An Assessment of Information-Based Warfare and U.S. National Security," by Charles Everett, Moss Dewindt, and Shane McDade, provides a retrospective and prospective review of information-based warfare in a national security context and within the context of the next revolution in military affairs. In a well-documented paper recognized in an honorable mention category, Colonel Brian Fredericks summarizes information warfare at the three-year mark with the admonition: Where do we go from here?

Winners of the 1995 award include Dr. John Miller's paper "Information Warfare: Issues and Perspectives," which reviews the elusiveness of the concept of information warfare. His discussion ranges from a narrowly defined context of information warfare focused on military operations to a much broader discussion of information warfare as an offshoot of the information revolution. In "A Chapter Not Yet Written," Colonel Adolph Carlson grounds his discussion of information management and the challenge of battle in case studies from the

Civil War (The Case of Fitz John Porter) and the Persian Gulf War (The Case of General Fredrick Franks) to illustrate the enduring issues of decision making under pressure in information-rich and -poor environments. Lastly, LTC Steven G. Fox, USA examines the "Unintended Consequences of Joint Digitization" including a discussion of the potential for merging the operational and tactical levels of war, diminishing a commander's prerogatives, and increasing the fragility of the force.

Papers selected for the Sun Tzu Award were selected using a formalized peer review process. Reviewers selected winning submissions on the basis of originality, innovativeness, and potential contribution to national security policy and strategy development. Particular emphasis was given to the grounding of the author's thesis in history and projecting implications for future conflict scenarios.

Please note that papers contained in this collection were submitted as part of an academic endeavor and should be viewed within an academic context. These essays represent the views of the individual authors and do not necessarily reflect the official opinion of the Information Resources Management College, the National Defense University, or the Department of Defense.

Special thanks are due to Dr. Daniel Kuehl of the School of Information Warfare and Strategy, NDU, for his continued efforts as a reviewer, and to Captain Gina Oliver, USAF, for her efforts to assemble this document into a coherent whole.

Dr. Robert E. Neilson
Editor

Sun Tzu and Information Warfare

⦾ KNOWLEDGE STRATEGIES:
Balancing Ends, Ways, and Means in the Information Age

Lieutenant Colonel William R. Fast
U.S. Army

ABSTRACT: Information age technologies are changing values and national interests, both of which drive the formulation of national security strategy. The *strategy equals ends plus ways plus means* paradigm must change. Information age knowledge strategy seeks the *ends* of cooperative and dynamic competition, uses the *ways* of network node control and organizational adaptation, and requires the resource *means* of valued information enhanced by experience in exploiting that information. A successful information age security strategy requires that we balance the *ends, ways, and means* of knowledge strategies. Whether we use the political, economic, military, or informational elements of national power, we serve our strategic ends best when we cooperate to shape robust information networks that promote dynamic competition and enhance mutual performance both in the public and private sectors. Further, we must control network nodes and communications links and secure our information resources. The security and integrity of our cyberspace must be considered an important, if not vital national interest.

Introduction

As we enter a new technological age, devising the proper national security strategy can have a profound effect on the outcome of war. There is no better example than the French approach after World War I. During the interwar period from 1919 to 1939, France formulated a weak and vulnerable strategy of forward defense, driven by her obsession with the methodical battle technique perfected at the end of World War I. On 10 May 1940, the world watched with horror as Germany, with far fewer resources, successfully invaded the Low Countries and Northern France. Germany had made the right strategic choices; her blitzkrieg concept of warfighting took full advantage of the mechanization of warfare.[1] While France was mired in an older strategy, Germany was energized by emerging technology to develop a bold offensive strategy.

Today, man's ways of making war are changing again because of new information age technologies. What can we do today to avoid repeating the French debacle? In *War and Anti-War*, Alvin and Heidi Toffler argue that we need to formulate a capstone concept of knowledge strategy to effectively take advantage of these information age technologies.[2] In other words, we need to understand the *ends, ways, and means* of information age strategy.

Change introduced by the information age is arguably greater than that which faced the post-World War I nations.[3] Moreover, knowledge strategy encompasses more than the military element of power. Knowledge strategy must also address the political and economic aspects of power, which become even more useful in the information age. Further, the extent to which we allow our organizational structures and social patterns to change will determine the success of knowledge strategy.

This paper describes the effects of information age technologies on our values and national interests, both of which drive the formulation of security strategy. It also explains how the *ends, ways, and means* paradigm of strategy must adapt to the emergence of information age warfare.[4] Finally, this analysis postulates a framework for formulating knowledge strategies.[5]

Changing Values and National Interests

Values

The information age brings a new level of personalization to our world that changes the value of consumer products and services. When ordering a new car, computer, or even new suit of clothes, we can customize the item to our needs, desires, and even our own physical measurements. While our personal buying habits have always characterized us as individuals, now the vendor can easily capture our unique preferences on bits of digital information. The value added to a product customized to personal preference is the value of knowledge. No longer do we have to accept the statistical norm.[6] We have come to expect and receive personalized products and services. We value personalization. Now the information-based market can tap this added value.

Americans also value their rights as individuals. The information age promotes and enhances these rights by empowering the individual. Unlike television and radio, information age digital communications allow on-demand programming—we simply have to ask explicitly for what we want and when we want it. With a computer terminal and telephone modem, an individual can trade shares any time of the day on any of the world's major stock exchanges. Telecommunications and virtual reality technologies make it possible for doctors at the Mayo Clinic to perform surgery on patients in any part of the world. In sum, the information age empowers individuals with access, mobility, and the ability to effect change anywhere, instantaneously. This is what makes the information age so different from the past.[7]

The value that we place on personalization and individual rights affects our world view and our expectations of nation-states. Single-issue politics forces our government to act on problems that are important to a few but often secondary to the majority. For example, the narrow interests of lobbyists have a disproportionate impact on legislation passed by the U.S. Congress. Knowledge workers, arguably better informed in their narrow fields of endeavor than government regulators, increasingly resent and even oppose government intervention.[8] They use the words *privatize*, *liberalize*, and *deregulate* when advocating the rules for applying information age technologies to businesses.[9] We must be careful not to politically disenfranchise these knowledge workers and their virtual communities.[10]

Spurred by information age technologies, our highly personalized social and political processes have become interconnected and nonlinear, making it difficult to distinguish cause from effect and effect from cause. We have created more nodes of power and influence. Our cyber-future will feature direct participation by the individual as opposed to group representation.[11] As a result, the relevance of authority and sovereignty have diminished.[12] This is not bad. In 1787, James Madison said: "To give information to people is the most certain and the most legitimate engine of government."[13] Yet harnessing the power of that engine is the challenge of knowledge strategy. How do we define national interests and objectives, the *ends* of strategy, in the information age?

National Interests

As its value increases within our global economy, information is fast becoming a strategic national asset. Natural resources (minerals, oil, etc.), long the strength of a growing industrial economy, are becoming less important. This is because information-based economies place more importance on intellectual capital and intellectual labor than on material capital and physical labor.[14] In addition, the computers that manipulate this information are potential first-strike targets. Most of our $6 trillion domestic economy depends upon our 125 million computers tied together by land- and satellite-based communications.[15] Protecting this infrastructure must now be considered as a primary security objective.

We have already witnessed the growth of national economic partnerships. An example is the partnership of American Airlines, MCI, and Citibank. Travel on American Airlines, phone calls on MCI, and charges on Citibank's credit card now earn free American Airlines trips for the user.[16] Through networking, the information age will allow more international economic alliances as well. The paradox is that networked economic alliances decrease the sovereignty of the nation-state. When the exchange of value occurs by electronic transmission rather than the transfer of products, trade policies become less important than the location of the network nodes.[17] Governments that take the lead in understanding and building networks will gain enormous comparative advantages.[18] Thus, pursuit of economic well-being and prosperity take on new prominence in the information age.

Similarly, the information age elevates the importance of political interests over security interests. Information age technologies can seriously erode totalitarian regimes. The political change in Central and Eastern Europe from 1989 to 1991 was not the aftermath of war, but the result of peaceful movements for individual rights, democracy, and better economic conditions.[19] Encouraging the nations and peoples of the world to value human rights and democratic principles becomes easier with the Internet and direct broadcast television. In addition, political alliances become easier to maintain as common understanding replaces chaotic misunderstanding. The Clinton administration has understood this shift. One objective of Clinton's National Security Strategy is the enlargement of the community of democratic states committed to free markets and respect for human rights.[20] Clearly, information age technologies are tools of preventive diplomacy; they can help promote democracy and human rights in those states where we have the greatest concerns for stability and security.

Thus, the information age has changed the nature of our economic and political interests and impacted on our national security interests. During the Cold War, concerns for power balance drove our economic policies and diplomatic relations. It was a zero-sum game. Trade sanctions, embargoes, and prohibitions on exporting critical wartime technologies severely distorted our economic policies. At times, we supported nations despite their politics or stand on human rights so long as they didn't embrace communism. Unlike the Cold War era, political and economic interdependency in the information age requires cooperation and the open exchange of knowledge.[21] We now play in a non-zero-sum game where win-win outcomes are not only expected but are required for democracies and information-based economies to flourish.

More than 2,300 years ago, the ancient Chinese strategist Sun Tzu appreciated values, interests, and the rational comparison of power. Before launching a military campaign, he said that the temple council should compare unity on the homefront and the morale of the army with that of the enemy. He also understood the inevitable economic burdens that war laid upon the people.[22] So it is today. Understanding shifts in our values, interests, and in the relative importance of the elements of power helps us understand why the *ends, ways, and means* paradigm of national security strategy must change in the information age.

Changing the Ends, Ways, and Means Model of Strategy

The *ends, ways, and means* paradigm posits that *strategy* equals *ends* plus *ways* plus *means*. *Ends* are expressed as national objectives drawn from national values and interests. *Ways* are courses of action to achieve *ends*. *Means* are the resources (manpower, materiel, money, forces, logistics, etc.) required to support each course of action. Unless *ends, ways, and means* are compatible and in balance, the strategy will be at risk. And the greater the imbalance, the greater the risk.[23] The information age changes all three elements of the strategy equation.

Ends

In the information age, national objectives (*ends*), other than the protection of the national information infrastructure, are not easily identifiable. Clearly, the emergence of global economic networks delink national corporations from national markets and turn them multinational. For example, profits from the sale of a new Boeing 777 aircraft find their way into countries worldwide. Boeing is a broker in the global economic network, buying materials and components worldwide, basing its acquisitions on price, availability, quality and any other number of factors. In effect, Boeing attempts to optimize its entire operation globally. As it does so, it pays little attention to national allegiance. In such an environment, governmental influence over Boeing's purchases becomes problematic. Then the implications of a power struggle between government and industry are very real.[24]

Economic security and prosperity in the information age are functions of a kind of equality between nations and firms. The more firms act internationally, as in the Boeing example, the less they can be held to national accountability. Walter Wriston asserts that "Capital will go where it is wanted and stay where it is well treated."[25] Multinational firms play one nation-state against the other as they seek the greatest profit.[26] Now trade agreements among sovereign nations are really inadequate when they don't include the concerns of global business organizations.[27] The North American Free Trade Agreement and the European Union are recent attempts of nations to achieve competitive equality with the growing multinational economic networks. Yet, in a global information age economy, it will be futile for sovereign states to attempt to cut off and control even part of the world market.[28] The organizing principles for the analysis of power have changed. Multinational firms anticipate events and react quickly in global markets. But governments, whose policies are geographically bound, react more slowly.[29] Power in the information age depends more on the ability to influence access and interconnection than on the capacity to enforce borders. It follows that the *ends* of our national security strategy will depend less on confrontation with opponents and more on cooperation and trust among competitors.[30] Moreover, total agreement on objectives within a globally linked network is virtually impossible.

If national economic objectives can't be achieved due to the emergent global and networked nature of markets, why not ignore global markets completely? Well, ignoring the networked global markets is risky business, if not impossible, for either a nation-state or a business concern. Each year since 1965, the U.S. commercial sector has invested more of its dollars in research and development than has the Department of Defense (DoD).[31] If our military services are to preserve their technological superiority over potential foes, they must have access to these commercial products. Similarly, individual businesses can afford neither the enormous costs nor bear the high risks of remaining on the leading edge of all information age technologies. Yet they can't afford to miss a breakthrough that could create new product lines. When businesses share intellectual capital (knowledge) through participation in global markets, they avoid isolation from new technologies.[32]

Obviously, the *ends* of our strategy equation have become unclear, since it may be difficult to achieve all desired national objectives in the globally networked information age. At best, a sovereign nation might effectively pursue its interests only as it paradoxically subordinates those interests to the common interests of all networked partners.[33]

Ways

It is not difficult to show how the *ways* of security strategy change with the information age. For example, information age

10

weapons are equalizers. They help small nations against large nations and favor the weak over the strong. Examples include Stinger missiles used by the Mujahedin against the Russians and computer viruses designed to invade individual weapon systems or an entire defense computer network.[34] However, the real problem lies in the fact that today's breakthrough technologies in electronics, computer systems, software, and telecommunications come from the commercial marketplace and are available to anyone in the world. Furthermore, foes may use these technologies to their advantage without even resorting to military applications.

In broadest terms, information warfare is not new. It encompasses any hostile activity directed against our knowledge and belief systems.[35] Cyberwar, the newest subset of information warfare, needs no battlefield—it is fought in cyberspace. Cyberspace includes information itself, the communication nets that move it, and the computers that make it useful.[36] Cyberspace can be influenced and at times dominated by anyone possessing inexpensive computers linked into existing global communication nets. The enemy may exploit global business organizations that produce cyber technology and determine the patterns of change.[37] He may attempt to propagate waves of data big enough to crash the network by overloading network switches.[38] Cyberwar operations can blind us electronically and may change the definition of what is a hostile attack and what determines defeat.[39]

Under the microscope of world opinion formed by means of pervasive communication satellites, open warfare is no longer an option for sovereign nations to pursue their national interests.[40] Cable News Network coverage can rapidly trigger a negative international response, as we have seen during the recent wars in Somalia and Bosnia. However, the information age offers a more subtle approach—waging a quiet war in cyberspace where digital fingerprints are hard, if not impossible, to trace.[41] When information warfare enters and uses public cyberspace, collateral damage may be significant. Banking, finance, telecommunications, trade, travel, energy, and cultural systems are vulnerable.[42] Misinformation and disinformation campaigns are easily mounted and hard to defend against. Moreover, an adequate defense depends upon gathering, analyzing, and distributing intelligence to a flexible, networked interagency team.[43]

So, the information age introduces at least three new concepts in the *ways* of strategy. First, information age weapons are equalizers and can negate the military principle of mass. Second, cyberwar needs no battlefield and therefore no specially trained military organization—even civilians may participate. Finally, the initial offensive strike in a quiet cyberwar would be hard to detect and to defend against. It is also impossible to limit the cyberwar battlespace to purely military networks.

Another way of assessing the changes in the *ways* of strategy is to compare World War I and II warfare to information age warfare. Whereas the world wars used attrition (WW I) and maneuver (WW II), information age war emphasizes control. Whereas the world wars attempted to exhaust (WW I) and annihilate (WW II), cyberwar seeks to paralyze. And whereas the tools of the world wars were firepower weapons (WW I) and mechanization (WW II) produced in mass, the tools of information war are limited numbers of inexpensive computers linked via global communication systems.[44]

Means

Knowledge as a resource is not included in the current resource paradigm of manpower, materiel, money, forces, and logistics.[45] Knowledge, the "ammunition" of information war, is inexhaustible. Once produced (at a cost), knowledge can be used repeatedly — it will not disappear. In fact, it only increases! Digital knowledge can be copied and never missed. It can be given away but still kept. Digital knowledge can be distributed instantly. It is non-linear; it defies the theory of economy of

scale.[46]   Knowledge is the key element of wealth in the
information age. Compared with industrial age manufacturing,
information-based industries can produce more with fewer
resources, less energy, and less labor. Production runs of *one* are
possible and even economical with intellectual capital
(knowledge) encoded in software and used by smart machines.[47]
The result is an explosion of personalized products and services.[48]
Moreover, knowledge to inform people, coded as digital bits,
can be turned into audio, video, or even graphics—it is
"mediumless."[49] Manpower, materiel, and forces, on the other
hand, possess *none* of these characteristics.

Knowledge as a resource is often cheaper than materiel. It
uses limited manpower or forces and may require little or no
logistics.   Thus the information age opens the doors to the
resource poor. Knowledge diffuses and redistributes power to
the weaker actors. It redraws boundaries and time and space
horizons. It enables organizations to open up.[50] When it comes
to balancing *means* with *ends* and *ways*, knowledge as a resource
offers an economical solution.

In sum, it is difficult to apply the *ends, ways, and means*
paradigm of strategy to information age security.   Unlike
traditional *means*, knowledge is relatively cheap and easy to
balance with *ends* and *ways*. Unlike conventional *ways*, cyberwar
defies the military principle of mass. And its primary objectives
are control and paralysis. Unlike the clearly articulated *ends* of
Cold War security strategies, national objectives in a globally
networked information age are more difficult to define and thus
to achieve. Clearly, we need a new framework for formulating
information age knowledge strategies.

A Framework for Formulating Knowledge Strategies

We can formulate knowledge strategies only with an
understanding of the strategic environment of the information
age. We can characterize this environment through three central
concepts: cooperative and dynamic competition, the wisdom

pyramid, and the productivity paradox. Also important is an understanding of how the bureaucracies of the industrial age might transform into the cyberocracies of the information age. Finally, we must review the importance of information dominance in cyberwar. This background and understanding will enable us to develop a formula for knowledge strategy.

Strategic Environment

In *Being Digital*, Nicholas Negroponte proclaims with optimism that "the control bits of that digital future are more than ever before in the hands of the young."[51] This is a profound statement when you consider the relatively advanced age of those who are currently responsible for formulating knowledge strategies! Fortunately, commercial knowledge industries are at the forefront in formulating knowledge strategies; they can enlighten us on the characteristics of the strategic information age environment.

The movement of portions of the silicon chip industry from Northern California to Bangalore, India, is an example of the environment knowledge industries create. Historically, innovative entrepreneurs in the Silicon Valley in California have made our computer chips. Now it seems that much of this chip design and engineering has moved to Bangalore. The reason: Bangalore engineers work for $500 per month, compared with $15,000 per month for an engineer in the Silicon Valley. Further, it is no coincidence that Bangalore is also the center of the Indian atomic energy industry. As American firms pour money for computer chips into Bangalore, one must ask what this investment is doing for India's nuclear weapons program? Clearly, the ability of our government (or the government of India) to control such economic activity at the national level is in steady decline as the entrepreneurial net draws the entire world more closely together.[52]

In a global information economy, the growth rates of individual countries should converge over time. As in the silicon chip example, India gains the newest, most innovative computer chips while U.S. firms absorb all the costs and risks.[53] Moreover, such alliances could create new free markets. Cybernations consisting of many like-minded virtual companies with cyber-economies could emerge. Cultures that have vanished from the real world may yet be reborn in cyberspace. A network superpower may emerge.[54] Thus, the strategic environment of the information age equalizes competitors while creating a potential for international instability.

Cooperative and Dynamic Competition

Another lesson of the silicon chip industry is that knowledge industries today seek cooperative competition, a framework that simultaneously enhances mutual performance but shapes the form of their competition. The United States could also pursue a strategy of cooperative competition in building global information age networks that would allow her to pursue her national objectives in concert with other nations. Most important, cooperative competition would allow us to shape the competition by controlling the protocols of these information networks.[55]

We can become a strategic network broker, balancing competition and cooperation with other nations by controlling access to and participation in these networks. As the strategic broker, we would have the upper hand in formulating the rules for competition. Yet the fact that we cooperate with the nations of the world promises them benefits such as converging growth rates. All nations could compete for the location of high value economic activities. Within the U.S., cooperative competition would promote a healthy domestic environment of technological and organizational innovation. Government policy would not stifle but encourage and support industry to reach out and tap knowledge banks throughout the world. In the information age, an alternate strategy of isolation supported by policies to shelter domestic industry (as experienced in the industrial age) could have disastrous consequences.[56]

Beyond cooperative competition, we also need *dynamic competition*: competition that allows new technologies to compete against and replace older technologies. In earlier times, dynamic competition gave us the automobile while the world was still looking for stronger horses (termed *static competition*). In the 1980s, dynamic competition transformed the computer industry from mainframes to mini and personal computers. It gave the U.S. world dominance in telecommunications, mircroelectronics, computer networking, and software applications. Significantly, American business and technological leadership created these vast new markets, not government oversight or policy making.[57] Through dynamic competition, we can further shape our competition and reap the greatest possible benefits from our information age economy.

Wisdom Pyramid

While the information age equalizes competitors, the wisdom pyramid mitigates against instability. Visualize a pyramid with the base composed of *raw data*. Add the next layer and call it *information* that rises like cream to the top of the data. On top of information, lay down another layer called *experience*. Finally, cap the pyramid with *wisdom*. Each person is a product of his or her own experience. Information, filtered up through that experience, creates wisdom at the top of the pyramid.[58] So it is with nation-states. The data and information others gain through information age networks has real value only as it filters through real experience. More important, corporate knowledge embedded in teams—like NASA's team that put man on the moon—is knowledge that none of the individual team members knows alone.[59] Embedded knowledge is hard, if not impossible, to steal. Thus our experience and social networks that develop and use information technologies are precious commodities. We can identify them as our strategic center of gravity in the information age environment.

Productivity Paradox

Another precept of the information age is that useful applications of knowledge require adaptive organizations and processes. The productivity paradox says that, initially, organizations will insert new information technologies into existing organizational structures. These technologies will simply improve the speed and increase the efficiency of current processes. However, to take full advantage of the technology, organizations need to change their processes and adapt their structures.[60] In this way, we tailor our knowledge to specific applications and capture the value of exchanged information.[61]

Information age military forces, evolving in their use of cyberspace, will follow the same path—first accommodating information technologies by incorporation, and next by reinventing their processes and adapting their organizational structures.[62] We see technological incorporation in the Army's effort to digitize the battlefield. The objective today is to add "applique" computers to combat vehicles to improve situational awareness. Yet true leveraging of computers depends less on improving situational awareness in every combat vehicle and more on how the entire combat force reconfigures itself to exploit the knowledge gained through the added technology. Such reinventing exploits the exponential power of information networks.

Success in future wars will require armed forces with open, adaptable organizations that can react more quickly to changes than can the competition.[63] These organizations must easily reconfigure to fill specific needs, saving time and money in the process. Such open organizations are not wedded to any one operating system; they can rapidly incorporate new information age technology. Ultimately, they must be adaptable to the knowledge they use.

Cyberocracy

The differences between a bureaucracy of the 20th century and a cyberocracy of the information age highlight the importance of organizational adaptation. Whereas bureaucracy forces and often limits information flow through defined channels connecting discrete points, cyberocracy broadcasts large volumes of information among many interested parties. Whereas bureaucracy emphasizes the hard quantitative skills of programming and budgeting (like DoD's Planning, Programming, Budgeting and Execution System), cyberocracy emphasizes soft skills such as policy management and understanding culture and public opinion. Whereas bureaucracy observes traditional boundaries between public and private sectors, cyberocracy breaks across these boundaries and allows for mixing of public and private interests. Bureaucracies must transform into cyberocracies if the new techniques of the information age are to take hold.[64]

A cyberocracy should have greater capability than a bureaucracy for dealing with the complex issues of an interconnected world. Yet to transform our organizations we must break the paradigm that establishes "big budgets" and "big staffs" as the basis of bureaucratic power. We must demonstrate the value of "big information" as the source of power in a cyberocracy.[65]

Information Dominance

In *Infotrends*, Jessica Keyes notes that "Most organizations suffer from a proliferation of data that is either redundant or underutilized. These same organizations suffer from not recognizing the true value of their data."[66] Once the value of data is understood, knowledge derived from that data can be used offensively to increase an edge or defensively to reduce an edge held by an opponent.[67] The ability to recognize the value of data and use this data to derive knowledge is the first step toward information dominance.

Information dominance is achieved by transforming knowledge into capability. It is the ability to identify the vulnerabilities and centers of gravity of an enemy, or even a competitor or customer. It is the capability to reshape organizations and revise strategies based upon a systematic analysis of the opponent.[68] For example, Federal Express (FedEx) won unchallenged leadership in global express delivery services when it realized "that information about the package is just as important as the package itself."[69] Understanding that the customer cares about where his or her package is at anytime, FedEx transformed its knowledge of bar coding, hand held computers, and global telecommunications into the capability to provide near real-time location information on every package in their possession.[70]

Knowledge–based alliances that share resources and save costs can also propel technology to new heights while preserving competition. For example, IBM and Apple Computer agreed in 1991 to share knowledge to create a new computer operating system based upon object-oriented technology and desktop multimedia software. Such a venture was too costly for just one company to undertake.[71] Recognizing strategic uses of information technology and leveraging intellectual capital, as in the cases of FedEx, IBM, and Apple Computer is truly in the realm of strategic art. However, as we found with the productivity paradox, such success comes through process and structural changes within the organization.[72]

At the national strategic level, we should build flexible organizations (cyberocracies) around information and intelligence processing, rather than around traditional functions and bureaucratic departments.[73] National information dominance is achieved through the fusion of all networks (similar to the fusion of human, signals, electronic, and other kinds of intelligence into *all source* intelligence). Offensively, national information networks can change the minds of our adversaries if they are synchronized to carry specific but coordinated messages.[74]

19

Defensively, a national information security strategy is required for the protection of our key information systems, to include their nodes, communications links, and data. The effort exceeds the responsibilities of the joint military services; critical information and networks belonging to all federal agencies, the private sector, and even our allies must as well be protected.

A Formula for Knowledge Strategy

To this point, we have identified several facets of the strategic information age environment and cybercratic institutions that shape knowledge strategies. Before redefining the strategy paradigm, we must recall two additional characteristics of network theory: First, value is added only at nodes; second, the strength of networks comes from their redundancy, or multiple pathways between any two points.

Consider our nation's interstate highway network and how it has enabled our economy to grow. Many businesses and industries locate close to city beltways (nodes) and bring great wealth to these areas. Moreover, when adverse weather or construction blocks one route, usually a near-by route can handle the traffic. Similarly, governments that take the lead in shaping information networks and in locating nodes within their borders stand to reap enormous comparative advantage.[75] Because of multiple nodes and pathways, networks have no center of gravity and must be defeated in detail.[76] Moreover, bureaucracies might be defeated by networks (cyberocracies), so it may take networks to counter other networks. "The future may belong to whoever masters the network form."[77]

With some modification to the meaning of the additive terms, knowledge strategy fits the *strategy equals ends plus ways plus means* equation. It follows from the discussions above that knowledge strategy (KS) seeks the *ends* of cooperative and dynamic competition (C/DC), uses the *ways* of node control and organizational adaptation (NC & OA), and requires the resource *means* of valued information (VI) enhanced by experience (E). Symbolically, the strategy equation changes to this:

Knowledge=Cooperative/Dynamic + Node Control & + Information
Strategy
        Competition   Org. Adaptation   Dominance

 KS   = (C/DC)   + (NC & OA)   +   (VI x E)
Strategy = *Ends*   + *Ways*   +   *Means*

Knowledge strategies focus on the strategic broker in crafting the rules of information networks. Cooperative and dynamic competition permits us to pursue our national security objectives in concert with other nations while shaping the competition. Control of network nodes adds value to information, strengthening information dominance and denying dominance by others. Organizational adaptation overcomes the productivity paradox and ensures that we exploit information networks to their fullest potential. Finally, knowledge strategies require information dominance that comes from the value of information enhanced by experience.

Knowledge strategies incur a degree of risk unless we balance all elements of the equation. Unbalanced conditions can result if cooperative and dynamic competition are not the stated objectives of the strategy, if we don't control the network nodes, if productivity suffers because the organization hasn't truly adapted to the technologies, or if the value of information is high but the experience to exploit this information is low.

Conclusion

The information age has shifted the focus of our values and national interests. Empowered by information age technologies, we have come to value individual preference in products and services and direct participation in the democratic process. Similarly, the pursuit of economic well-being and the promotion of democratic values takes on added importance in contrast to our traditional security interests. Information is fast becoming a strategic national asset. Thus, the security and integrity of our cyberspace must now be considered an important, if not vital, national interest that we cannot afford to compromise.

A successful information age security strategy requires that we balance the *ends, ways,* and *means* of knowledge strategies. Whether we use the political, economic, military, or informational elements of national power, we serve our strategic *ends* best when we cooperate to shape robust information networks that promote dynamic competition and enhance mutual performance. Ironically, global information networks, built to bring peace and prosperity to the world, will be among the first attacked in a cyberwar. Denying access to these networks in hopes of preempting attack is totally counterproductive: it accomplishes the adversary's mission for him! Therefore, before an enemy attempts to fire the first hostile bits across our networks, we must control network nodes and communications links and secure our information resources.

Successful knowledge strategies require the mastery of information networks. Information networks operate on the win-win philosophy: one wins only if all win. The more our national interests reflect those of the networks, the better chance we have of achieving them. Thus, we must be the primary architects of networks and seek to broker network operations. At times, we must be willing to subordinate our national objectives to the greater objectives of the networked nations and multinational firms with whom we interact. We must be willing to share knowledge resources and enter into knowledge-based alliances that allow us to leverage information age technologies. Our government can empower information age enterprise and encourage innovation by easing access to global networks. In relations with other nations, we should trade economic network integration for democratic and human rights reform. A more stable and safer world is one whose players share similar values and interests and who depend upon each other in a globally networked market economy.

We must realize that our strategic center of gravity is shifting to encompass our experience and the virtual communities we establish to exploit the information environment. We must care for our knowledge workers and educate the youth of our nation who will take their place. We can't exploit the information age without them. The "hub of all power and movement" in the information age will be our dominant knowledge.[78]   Only through non traditional open organizations with decentralized power structures can we truly achieve this dominance. We must create cooperative cyberocracies organized around the knowledge workers and processes that can best exploit all available information networks. Thus the *ways* of a knowledge strategy must break down the boundaries between government bureaucracies and the private sector. Most important, the extent of organizational adaptation—and how much it ultimately transforms the rules of information age networks and cyberwar—will determine whether we are using information age technologies to our fullest advantage.

Finally, to resource information age strategies, we must recognize that knowledge is a very economical *means* that can stretch and positively leverage our nation's wealth. Declining defense budgets have been—and will continue to be—the primary engines transforming the U.S. military and driving information age technologies into the hands of our armed forces. However, just as our armed forces engage in a revolution in military affairs, so must other government agencies and the private sector engage in revolutions in political, economic, and informational affairs.   Big bureaucracies with big operating budgets must downsize and leverage the power of information. We must share knowledge resources within the federal government and between the public and private sectors, even as they are transforming to adapt to the information age. We must invest only in those information age technologies and intellectual capital that will generate the most significant returns in information dominance.

Again, we should recall that France's disappointment in World War II was not that she was surprised, but that she made the *wrong* strategic security choices.[79] France knew that war with Germany was coming. So she prepared for that war. However, she failed to understand the significance of the new mechanized age. Germany understood the strategic importance of mechanization and overwhelmed France with the blitzkrieg.

So it is with the United States today. The dawning information age gives us an opportunity to make strategic choices. We must not simply continue the security strategies of the past. Rather, we must seek to understand the strategic importance of knowledge and discover the rules of cyberspace and cyberwar. Understanding how to balance the *ends, ways,* and *means* of knowledge strategies is the first step in making the *right* strategic choices for the emerging information age.

Notes

1. Robert A. Doughty, *The Seeds of Disaster: The Development of French Army Doctrine 1919-1939* (Hamden, CT: Archon Book, 1985), 190; Gordan R. Sullivan and Anthony M. Coroalles, *Seeing the Elephant: Leading America's Army into the Twenty-First Century* (Hollis, NH: Puritan, 1995), 20.

2. Alvin Toffler and Heidi Toffler, *War and Anti-War: Survival at the Dawn of the 21st Century* (Boston: Little, Brown & Co., 1993), 141.

3. Sullivan and Coroalles, 21.

4Arthur F. Lykke, Jr., ed., *Military Strategy: Theory and Application* (Carlisle, PA: U.S. Army War College, 1993), 3.

5. Richard A. Chilcoat, *Strategic Art: The New Discipline for 21st Century Leaders* (Carlisle, PA: U.S. Army War College, 1995), 4.

6. Nicholas P. Negroponte, *Being Digital* (New York: Knopf, 1995), 155 and 164.

7. Ibid., 166, 169 and 231.

8. Walter B. Wriston, *The Twilight of Sovereignty: How the Information Revolution is Transforming Our World* (New York: Scribner's, 1992), 38.

9. Richardo Petrella, "Information Society - Future Prospects" (British Computer Society and UNISYS Annual Lecture, 4 Nov 94); available from http://www.bcs.org.uk/unisys.html; Internet; accessed 14 Apr 96.

10. Joe Costello, "Electronic Polity" (Quantum Polis, 17 Feb 95); available from http://www.cts.com/browse/joec/qpolis/local.html; Internet; accessed 14 Apr 96.

11. Joe Costello, "The Organic Information Age" (Quantum Polis, 21 Jun 94); available from http://www.cts.com/browse/joec/qpolis/oia.html; Internet; accessed 14 Apr 96.

12. Donald N. Michael, "Too Much of a Good Thing? Dilemmas of an Information Society," *Vital Speeches of the Day* L, no. 2 (1 Nov 83): 54.

13. John P. Foley, ed., *The Jeffersonian Cyclopedia* (New York: Funk and Wagnels, 1900), 277.

14. Wriston, 6.

15. Winn Schwartau, *Information Warfare: Chaos on the Electronic Superhighway* (New York: Thunder's Mouth Press, 1994), 12-13.

16. Jessica Keyes, *Infotrends: The Competitive Use of Information* (New York: McGraw-Hill, 1993), 200.

17. James R. Golden, *Economics and National Strategy in the Information Age: Global Networks, Technology Policy, and Cooperative Competition* (Westport, CT: Praeger, 1994), 16-17.

18. David F. Ronfeldt, *Cyberocracy, Cyberspace, and Cyberology: Political Effects of the Information Revolution* (Santa Monica, CA: RAND, 1991), 77-78.

19. Sir Leon Brittan, "EU Pursues Global Answers: International Economic Instability is New Threat," *Defense News* 10, no. 48 (4 Dec 95): 27.

20. The White House, *A National Security Strategy of Engagement and Enlargement* (Washington: February 1995), 22.

21. Golden, 7.

22. Sun Tzu, *The Art of War*, trans. Samuel B. Griffith (New York: Oxford University Press, 1971), 39-40 and 63-71.

23. Lykke, 3-5.

24. Jorge Reina Schement and Terry Curtis, *Tendencies and Tensions of the Information Age* (New Brunswick, NJ: Transaction, 1995), 63.

25. Wriston, 61.

26. Joe Costello, "In Defense of Progressives)" (Quantum Polis, 19 Dec 94); available from http://www.cts.com/browse/joec/qpolis/prog.html; Internet; accessed 14 Apr 96.

27. Schement, 64.

28. Wriston, 37.

29. Jerome Glenn, "Japan: Cultural Power of the Future," *The Nikkei Weekly* (7 Dec 92).

30. Golden, 16-17.

31. Paul G. Kaminski, "Investing in Tomorrow's Technology Today," *Defense Issues* 10, no. 46 (1995): 2.

32. Golden, 16.

33. Michael, 41; Wriston, 37.

34. Wriston, 15.

35. Richard Szafranski, "A Theory of Information War: Preparing for 2020" (n.d.); available from http://www.cdsar.af.mil/apj/szfran.html; Internet; accessed 14 Apr 96.

36. G.A. (Jay) Keyworth II, "Telecommunications: More Computing than Communications" (The Progress & Freedom Foundation, 31 Jan 95); available from http://www.pff.org/pff/Telecommunications.html; Internet; accessed 14 Apr 96.

37. Golden, xiv.

38. Ian D. Pearson, *The Future of Social Technology*, 60 min. (Atlanta: World Future Society, 1995), audiocassette and vu-graphs, 19.

39. John Arquilla and David Ronfeldt, "Cyber War Is Coming!," *Comparative Strategy* 12 (April-June 1993): 157; available from http://www.stl.nps.navy. mil/c4i/cyberwar.html; Internet; accessed 14 Apr 96.

40. Wriston, 13-14.

41. Martin C. Libicki, *The Mesh and the Net: Speculations on Armed Conflict in a Time of Free Silicon* (Washington: National Defense University, 1994), 26.

42. Szafranski, 5.

43. Thomas X. Hammes, "The Evolution of War: The Fourth Generation," *Marine Corps Gazette* 78, no. 9 (September 1994): 44.

44. John Arquilla, "The Strategic Implications of Information Dominance," *Strategic Review* 22 (Summer 1994): 26.

45. Lykke, 4.

46. Jerome Glenn and John Peterson, *Information Warfare, Cyber Warfare, Perception Warfare and their Prevention*, 60 min. (Atlanta: World Future Society, 1995), audiocassette.

47. Jerry Harris and Carl Davidson, "The Cybernetic Revolution and the Crisis of Capitalism" (The Chicago Third Wave Study Group, 18 Jul 94); available from http: //www. bradley.edu/ las/soc/ syl/391/ papers/cyb_revo.html; Internet; accessed 14 Apr 96.

48. Alvin Toffler and Heidi Toffler, *Creating a New Civilization: The Politics of the Third Wave* (Atlanta: Turner, 1995), 37.

49. Negroponte, 72.

50. Arquilla and Ronfeldt, 142.

51. Negroponte, 231.

52. Charles William Maynes, "The World in the Year 2000: Prospects for Order or Disorder," *The Nature of the Post-Cold War World* (Carlisle, PA: U.S. Army War College, 1993), 21-22; Harris, 8; Pearson, 50.

53. Golden, 13.

54. Pearson, 43; Glenn.

55. Golden, 4.

56. Ibid., xv.

57. Esther Dyson et al., "Cyberspace and the American Dream: A Magna Carta for the Knowledge Age" (The Progress & Freedom Foundation, release 1.2, 22 Aug 94); available from http://www.pff.org/pff/position.html; Internet; accessed 14 Apr 96.

58. Wriston, 176.

59. Seev Neumann, *Strategic Information Systems: Competition Through Information Technologies* (New York: Macmillan, 1994), 109.

60. Ronfeldt, 9-10.

61. Golden, xiv.

62. Libicki, 3.

63. John J. Patrick, *Reflections on the Revolution in Military Affairs* (Washington: Techmatics, 1994), 8-9.

64. Ronfeldt, 21.

65. Ibid.

66. Keyes, 150.

67. Ibid., 188-189.

68. Arquilla, 29.

69. Fred Smith, founder of FedEx, quoted in Keyes, 194.

70. Ibid., 192-197
71. Neumann, 107-108.
72. Keyes, 198-199.
73. Stuart Johnson and Martin Libicki, eds., *Dominant Battlespace Awareness* (book on-line)(Washington: NDU Press, 1995), 30; available from http:// 198.80.36.91/ndu/inss/books/dbk/dbk1.html; Internet, accessed 14 Apr 96.
74. Hammes, 44.
75. Ronfeldt, 77-78.
76. Libicki, 33.
77. Arquilla and Ronfeldt, 155.
78. Clausewitz defines *center of gravity* as "the hub of all power and movement"; see Carl von Clausewitz, *On War*, ed. and trans. Michael Howard and Peter Paret (Princeton, NJ: Princeton University Press, 1976), 595.
79. Doughty, 190.

Bibliography

Arquilla, John and David Ronfeldt. "Cyber War Is Coming!" *Comparative Strategy* 12 (April-June 1993): 141-165. Available from http: //www.stl.nps.navy.mil/c4i/cyberwar.html; Internet; accessed 14 Apr 96.

Arquilla, John. "The Strategic Implications of Information Dominance." *Strategic Review* 22 (Summer 1994): 24-30.

Brittan, Sir Leon. "EU Pursues Global Answers: International Economic Instability is New Threat." *Defense News* 10, no. 48 (4 Dec 95): 27.

Chilcoat, Richard A. *Strategic Art: The New Discipline for 21st Century Leaders.* Carlisle, PA: U.S. Army War College, 1995.

Clausewitz, Carl von. *On War.* Translated and edited by Michael Howard and Peter Paret. Princeton, NJ: Princeton University Press, 1976.

Costello, Joe. "In Defense of Progress(ives)." Quantum Polis, 19 Dec 9 4 . Available from http://www.cts.com/browse/joec/qpolis/prog.html; Internet; accessed 14 Apr 96.

_____. "Electronic Polity." Quantum Polis, 17 Feb 95. Available from http://www.cts.com/browse/joec/qpolis/local.html; Internet; accessed 14 Apr 96.

_____. "The Organic Information Age." Quantum Polis, 21 Jun 94. Available from http://www.cts.com/browse/joec/qpolis/oia.html; Internet; accessed 14 Apr 96.

Doughty, Robert A. *The Seeds of Disaster: The Development of French Army Doctrine 1919-1939*. Hamden, CT: Archon Book, 1985.

Dyson, Esther, George Gilder, George Keyworth, and Alvin Toffler. "Cyberspace and the American Dream: A Magna Carta for the Knowledge Age." The Progress & Freedom Foundation, release 1.2, 22 Aug 94. Available from http://www.pff.org/pff/position.html; Internet; accessed 14 Apr 96.

Foley, John P., ed. *The Jeffersonian Cyclopedia*. New York: Funk and Wagnels, 1900.

Glenn, Jerome and John Peterson. *Information Warfare, Cyber Warfare, Perception Warfare and their Prevention*. 60 min. Atlanta: World Future Society, 1995. Audiocassette.

Glenn, Jerome C. "Japan: Cultural Power of the Future." *The Nikkei Weekly* (7 Dec 92).

Golden, James R. *Economics and National Strategy in the Information Age: Global Networks, Technology Policy, and Cooperative Competition*. Westport, CT: Praeger, 1994.

Hammes, Thomas X. "The Evolution of War: The Fourth Generation." *Marine Corps Gazette* 78, no. 9 (September 1994): 35-44.

Harris, Jerry and Carl Davidson. "The Cybernetic Revolution and the Crisis of Capitalism." The Chicago Third Wave Study Group, 18 Jul 94. Available from http://www. bradley.edu/las/soc/syl/391/papers/cyb_revo.html; Internet; accessed 14 Apr 96.

Johnson, Stuart and Martin Libicki, eds. *Dominant Battlespace Awareness*. Book on-line. Washington: NDU Press, 1995. Available from http://198.80.36.91/ndu/inss/books/dbk/dbk1.html; Internet; accessed 16 Apr 96.

Kaminski, Paul G. "Investing in Tomorrow's Technology Today." *Defense Issues* 10, no. 46 (1995): 1-17.

Keyes, Jessica. *Infotrends: The Competitive Use of Information*. New York: McGraw-Hill, 1993.

Keyworth, G.A. (Jay) II. "Telecommunications: More Computing than Communications." The Progress & Freedom Foundation, 31 Jan 95. Available from http://www.pff.org/pff/Telecommunications.html; Internet; accessed 14 Apr 96.

Libicki, Martin C. *The Mesh and the Net: Speculations on Armed Conflict in a Time of Free Silicon.* Washington: National Defense University, 1994.

Lykke, Arthur F., Jr., ed. *Military Strategy: Theory and Application.* Carlisle, PA: U.S. Army War College, 1993.

Maynes, Charles William. "The World in the Year 2000: Prospects for Order or Disorder." *The Nature of the Post-Cold War World.* Carlisle, PA: U.S. Army War College, 1993.

Michael, Donald N. "Too Much of a Good Thing? Dilemmas of an Information Society." *Vital Speeches of the Day* L, no. 2 (1 Nov 83): 38-42.

Negroponte, Nicholas P. *Being Digital.* New York: Knopf, 1995.

Neumann, Seev. *Strategic Information Systems: Competition Through Information Technologies.* New York: Macmillan, 1994.

Patrick, John J. *Reflections on the Revolution in Military Affairs.* Washington: Techmatics, 1994.

Pearson, Ian D. *The Future of Social Technology.* 60 min. Atlanta: World Future Society, 1995. Audiocassette.

Petrella, Richard. "Information Society - Future Prospects." British Computer Society and UNISYS Annual Lecture, 4 Nov 95. Available from http://www.bcs.org.uk/unisys.html; Internet; accessed 14 Apr 96.

Ronfeldt, David F. *Cyberocracy, Cyberspace, and Cyberology: Political Effects of the Information Revolution.* Santa Monica, CA: RAND, 1991.

Schement, Jorge Reina and Terry Curtis. *Tendencies and Tensions of the Information Age.* New Brunswick, NJ: Transaction, 1995.

Schwartau, Winn. *Information Warfare: Chaos on the Electronic Superhighway.* New York: Thunder's Mouth Press, 1994.

Sullivan, Gordan R. and Anthony M. Coroalles. *Seeing the Elephant: Leading America's Army into the Twenty-First Century.* Hollis, NH: Puritan, 1995.

Sun Tzu. *The Art of War.* Translated by Samuel B. Griffith. New York: Oxford University Press, 1971.

Szafranski, Richard. "A Theory of Information War: Preparing for 2020." n.d. Available from http://www.cdsar.af.mil/apj/szfran.html; Internet; accessed 14 Apr 96.

Toffler, Alvin and Heidi Toffler. *Creating a New Civilization: The Politics of the Third Wave*. Atlanta: Turner, 1995.

_____. *War and Anti-War: Survival at the Dawn of the 21st Century*. Boston: Little, Brown & Co., 1993.

The White House. *A National Security Strategy of Engagement and Enlargement*. Washington: February 1995.

Wriston, Walter B. *The Twilight of Sovereignty: How the Information Revolution is Transforming Our World*. New York: Scribner's, 1992.

◯ THE SILICON SPEAR: Assessment Of Information Based Warfare and U.S. National Security

Charles B. Everett, Moss Dewindt, & Shane McDade

> One very important reason for disliking a weapon was, of course, because it was new. A weapon might or might not be effective, but whenever one was introduced it always threatened to upset traditional ideas as to how war should be waged, and, indeed, what it was all about.
>
> —Martin Van Creveld
>
> *The Transformation of War*

The Setting

The First Battles in the Era of Information Based Warfare: The Seizure of Fiery Reef and Mischief Island: July 1997

In retrospect, it was all quite foreseeable. But then, hindsight is always 20/20. The events had been lost in the "noise." The Board of Inquiry and the numerous congressional investigations had all come to the same conclusion.

Of greater concern, however, was the fact that the U.S. Navy had wargamed similar events in 1994 at the Naval War College.

There, a resurgent Chinese military had badly bloodied U.S. forces. The reasons were quite evident—the PRC had a 21st century military while the U.S. had fielded an updated version of its Gulf War forces; and, of greater importance, the Chinese had understood early on that "zhan zheng xiang tai"—a change in the form of war—had taken place. They seized upon the concept of Information Based Warfare and melded it into their way of thinking and consequently set forth doctrine and strategy for "Bin Fa"—military tactics.[1]

The 1993 publication of a book entitled, *Can the Chinese Army Win the Next War?*, had not been taken seriously by U.S. policy makers. Caught up in events in Bosnia, Central Africa and election year rhetoric, the thrust of the book—the invasion of Taiwan, the seizure of the Spratlys and the Paracel Islands, with the United States China's principal military adversary—had been lost in the presidential campaign.

As the date for the transfer of Hong Kong neared in July 1997, the world was focused upon the increasingly belligerent actions of the Beijing Government. Refugees had begun leaving the island colony, en masse, in early June. World-wide concern was heightened by a series of international financial crises as financiers attempted to compensate for the outflow of money as the shadow government in Bejiing—still crippled by the long-anticipated death of Deng Xiaoping—sought to shape policy with the result being a series of pronouncements and actions that left western analysts even more confused.

Already stretched thin by a sophomoric national security policy that optimistically called for the U.S. military to concurrently cope with two major regional crisis (MRC), the U.S. Pacific Fleet was ill-prepared for little more than a show of force along the Asian mainland. Still wary of PRC military exercises in December 1995 and the pre-election show of force in the Straits of Taiwan during March of 1996, the National Command Center was operating at a heightened state of readiness. But, as has been the case historically, operating in

DEFCON II for an extended period of time left the national nerve center fatigued. What would have been recognizable to alert, rested analysts, was lost by those who had been working 12 hour shifts for almost 45 days.

On 18 July 1997, the President informed the American people that relations with the People's Republic of China were tenuous at best. This announcement was driven by SIGINT intercepts that revealed heightened military activities in the PRC. A U.S. carrier battle group, steaming in the north Pacific reported that it was being shadowed by several PRC submarines. The U.S. military attaché in New Delhi reported that PRC long range aircraft had overflown Indian airspace on three occasions. Reports coming out of one of the few news services remaining in Hong Kong noted that the 2nd Artillery—the PRC's nuclear rocket force—had begun to disperse firing battery's well south of Lop Nor, near the headwaters of the Mekong River in terrain that might preclude the travel of U.S. cruise missiles through the rugged Himalayas.

Definition

Information-based warfare is an approach to armed conflict focusing on the management and use of information in all its forms and at elf fevers to achieve a decisive military advantage in especially in the joint and combined environment. Information based-warfare is both offensive and defensive in nature—ranging from measures that prohibit the enemy from exploring information to corresponding measures to assure the integrity, availability, and interoperability of friendly information assets.

While ultimately military in nature, IBW is also waged in political, economic, and social arenas and is applicable over the entire national security continuum from peace to war and from 'tooth to tail.' Finally, Information Based Warfare focuses on the command and control needs of the commander by employing

state of the art information technology such as synthetic
environments to dominate the battlefield.

> —Working definition recognized by the
> School of Information Warfare of the
> National Defense University as of 16 Nov
> 96.

It is most appropriate that the NDU sponsored contest to
encourage the study of information based warfare is named for
Sun-tzu, a personage who has received almost god-like reverence
by those who would become students of the military art in the
west. However, the significance of the spirit of Sun-tzu should
be balanced by the significance of the similarities between the
ancient states of Ch'i, Chin, and Ch'in—which would eventually
give name to what is called China—and the United States. What
is clear from a reading of the *Seven Military Classics of China* is
the concern throughout the seven books for information that
would enable the rulers to have knowledge of their vast domains
and the enemies that posed threats to the Celestial Kingdom.
From the time of the legendary Sage Emperors (2852-2255 BC.)
through the Hsia, Shang, and Chou Eras, and beyond to the Chi'
in and early and late Han Dynasties, it was clear that information
was the basis for decisions on maintaining the peace and waging
war. The size of the Celestial Kingdom was simply too great to
launch an army whenever a potential threat loomed on the
horizon. Thus, it was that information based warfare—colored
with a distinctive Chinese flavor—came into being. For
westerners, the works of Sun-tzu best portray the seemingly
anti-western concept that the general who wages war without
engaging in actual conflict is the superior tactician.

The United States shares a similar legacy with early China.
As the world superpower it must have information from around
the world upon which to base its' policies and strategies—in
effect no different than Chang Liang's search for the information
that enabled him to establish power and consolidate the authority

of the Han dynasty.[4] Like ancient China, the United States today has not the resources to sally forth at the first sound of trouble. Like China, the U.S. must develop a strategy by which information can be acquired, evaluated, and acted upon— using methods short of war as the weapons of first choice. The concept of Information Based

Warfare has become a necessity for the U.S. as it attempts to define and protect its national security interests over a world that makes the Celestial Kingdom appear small by comparison.

Today, controversy rages in the U.S. military establishment as to both definition and application of Information Warfare. The current Joint Staff definition of Information Warfare is: "Actions taken to achieve information superiority in support of national military strategy by affecting adversary information and information systems while leveraging and protecting our information and information systems."

It is our belief that this definition highlights only the broadest character of IW and as such is far too abstract, depending on a narrow strategic environment which is inadequate when assessing the future conflicts that our national security strategy must address. This is not to assert, however, that little has been accomplished in attempting to better understand and appreciate the intricacies of IW. On the contrary, there exists a vast literature on the subject which spans from the civilian sector to the highest levels of the Intelligence Community and the Department of Defense.

While the literature concerning IW is substantial, we believe that there still exist gaps in its overall conceptual and theoretical framework. For purposes of clarity we have chosen to use the more descriptive—and conceptually more correct and advanced concept—of Information Based Warfare as opposed to Information Warfare. We believe IBW more accurately represents the "zhan zheng xing tad"—change in the form of war—that is currently being addressed by those who wish to understand the reality of future conflict.[5]

Not surprisingly, it is the ancestors of Sun-tzu, Sun Ping, Ssu-ma, and Wei Liso-tzu who appeared to have also seized upon the concept of Information Based Warfare and moved it beyond the present level of U.S. debate. As a potential adversary, the United States must quickly step up to the next theoretical plateau. To that end, we propose to trace the evolution of Information Warfare—a concept that we believe is second order in scope—to the next logical plateau of Information Based Warfare.

The Evolution of Information Based Warfare

Therefore, at times of revolution, when the normal-scientific tradition changes, the scientist's perception of his environment must be re-educated—in some familiar situations he must learn to see a new gestalt.

—Thomas S. Kuhn
The Structure of Scientific Revolutions

Whereas we had available for immediate purpose one hundred and forty-nine first-class warships, we now   have two, these two being the *Warrior* and her sister *Ironside*. There is not now a ship in the English navy apart from these two that it would not be madness to trust in an engagement with that little [American] *Monitor*..

—The Times (London), 1862

The most recent steps in that evolution can be found in the Military Technical Revolution (MTR) which may be attributed to Soviet Military thinkers; and, a recurrence of what is called the Revolution in Military Affairs (RMA).

In the early 1980's the Soviets noted that "the emergence of advanced non-nuclear technologies was engendering a new revolution in military affairs. They were particularly interested in the "incorporation of information sciences into the military sphere" and in the idea of a "reconnaissance-strike complex. The

38

events in the Gulf War convinced them of their hypothesis (comparative strategies). RMAs matter principally for two reasons. First, being second best may lead to catastrophic loss in future wars. Since the only objective benchmark for determining the relative effectiveness of forces (that is, success in combat) is unavailable in long periods of peace, there is great potential for asymmetries in combat effectiveness between militaries, observable only when the next war occurs. Secondly, as equipment life cycles, especially for platforms, steadily grow to encompass decades, many of the principal weapons systems of 2025 will likely be designed and built in the next few years. Since militaries are stuck with force structures they choose for long periods, it is more crucial than ever to think about them now, in peacetime, about the revolutionary changes in the nature of war and the about what will matter in winning wars in twenty or thirty years. Today with the United States arguably the only superpower for the foreseeable future, one might ask why this issue is especially pressing. Replicating the U.S. force structure is clearly beyond the reach of all but a few other nations, even in the long term. This may not be relevant. Even small-to medium sized powers may be able to exploit specific technologies for significant military leverage in certain areas. The current rate of change suggests that state of the art in any technological context will be an extremely short-lived phenomenon, particularly with respect to the technologies that were key to the success of Desert Storm: space systems, telecommunication systems, computer architecture's, global information distribution networks, and navigation systems. Future revolutions will occur much more rapidly, offering far less time for adaptation to newmethods of warfare. The growing imperative in the business world for rapid response to changing conditions in order to survive in an intensely competitive environment is surely instructive for military affairs.[6]

The present RMA entails a fundamental change in who, how, and, perhaps even why wars are fought. It is driven not only by new technologies but by new operational concepts, new tactics, and new organizational structures. The impact of the current confluence of social, political, economic, and technological forces on American society and armed forces may equal—or exceed—what occurred during the 1960's and 1970's during the turmoil associated with the war in Vietnam. [7]

The question that must be asked as we attempt to understand this RMA, is what were the results of change wrought by past Revolutions in Military Affairs? It has been suggested that accelerated interservice rivalries and over-reliance on management systems marked the last RMA, driven by the advent of atomic weapons at the end of WW II and the relatively stable and sparse defense budgets of the 1950's. [8] The current RMA is characterized by four types of changes:

- extremely precise;
- stand-off strikes;
- dramatically improved command, control, and intelligence;
- information warfare; and
- nonlethality

Many analysts see a number of benefits from harnessing the current revolution in military affairs and using it to build 21st century U.S. armed forces:

- rejuvenating the political utility of military power;
- delaying the emergence of a peer competition;
- providing a blueprint for technology acquisition and force reorganization;
- and inspiring conceptual, forward-looking thinking.

Looking ahead, many believe that the current RMA will have at least two stages. The first is based on stand-off platforms, stealth, precision, information domination, improved communications, computers, global positioning systems, digitization, "smart" weapons systems, jointness, and use of ad hoc coalitions. The second may be based on robotics, nonlethality, pyscho-technology, cyberdefense, nanotechnology, "brilliant" weapons systems, hyperflexible organizations, and "fire ant warfare." [9]

Steven Metz and James Kievit of the Strategic Studies Institute argue that a cost /benefit analysis of the present RMA needs to be pursued. One the one hand, they argue that a case can be made that costs and risks of vigorous pursuit of the current RMA outweigh the expected benefits. These include the risk that:

- the current RMA will not generate increased combat effectiveness against the most likely or most dangerous future opponents;
- American pursuit of the RMA will encourage opponents or potential opponents to seek countermeasures;
- the current RMA might lead the United States toward over reliance on military power; and, vigorous pursuit of the current RMA might increase problems with friends and allies.

On the other hand, they argue that there are very pressing reasons for supporting the current RMA:

> it should bring significant increase in combat effectiveness against some mid-level opponents; a force built around stand-off, precision weapons and disruptive information warfare capabilities would be more politically usable than a traditional force-projection military; the RMA could augment deterrence; and, the United States may need to pursue the current RMA to avoid stumbling into strategic inferiority. [10]

Lastly, they posit—and we strongly agree—that if policymakers decide to pursue the present revolution in military affairs, strategy, rather than technological capability should guide force development. The key question is: What do we want the future U.S. military to be able to do?[11] [We would go one step further and suggest that the RMA must lead toward answers to the question that is really the essence of strategy: How do we win?]

Unfortunately one finds little discussion of the foregoing ideas in the DOD's recently completed "Bottom-Up" Review. Former Secretary of Defense Aspin initiated the review with the laudable goal of rethink the basis for U.S. defense planning. It placed emphasis in many of the right areas: readiness, keeping forces for more than one regional war, acquisition reform. This review built a substantial consensus in the Pentagon behind the new force structure. Nonetheless, the review offers a classic example of military leaders planning to fight the last war. The report's proposed force for a single regional contingency—four to five Army divisions, four to five Marine brigades, 10 Air Force fighter wings, 100 heavy bombers, and four to five carriers—mirrors almost exactly the forces deployed in Operation Desert Storm. The review offers few thoughts on new technologies or techniques that might change the nature of war in coming decades. [12]

Clearly, the challenge is to move beyond the last war—an undertaking that is fraught with peril and promise.

An Information Based Warfare Model

We are pilgrims, Master; we shall go
Always a little further...."
—Inscription on the Clock, Bradbury Lines, Hereford,
Home of the Special Air Service Regiment
J. F. Flecker

Today

We believe that it is instructive to review the Program of Analysis for FY 1996 as set forth by the Directorate for Combat Support, National Military Intelligence Production Center, Defense Intelligence Agency. Entitled, *Intelligence Support to Infrastructure Warfare*, the thrust of this paper is directed toward a new paradigm that is centered on the principal of attacking an adversary's infrastructure to degrade or deny mobility, support to combatants and leadership. The paper sets forth valid arguments to support the thesis that to a large extent, recent advances in weapons and tactics were driven by the strategic objective of penetrating and rupturing the physical and psychological "centers of gravity" of the enemy warfighting capability. It is clear that new technologies have spurred an entire new generation of warfighting capabilities and, of equal importance, fundamentally altered the construct and performance of the modern nation-state. Discrete centers of gravity are giving way to diversification and interdependence in the modern nation-state. Information is the raw material that fuels productivity and power. Critical systemic nodes in the dynamic flow of commodities—consumables, services, and information—are at the same time increasingly obscure, strategically important, and tactically vulnerable. Fine grain analysis [read Intelligence writ in large script] and precision targeting fin both the conventional and non conventional sense] of these nodes and their synergistic dependencies is the centerpiece of Infrastructure Warfare. [13]

The primary elements of Infrastructure Warfare are:

- The nation-state as a System of Systems;
- Information as an Instrument of Power;
- Denial and Deception;
- Urban Infrastructure-The Operational Environment.

While the concept of Infrastructure Warfare is certainly a step in the right direction, we believe that this concept has several shortcomings that need to be addressed in order that it fully lead toward an answer to the strategic "bottom-line"—how do we win? We argue that the conflicts of the new world disorder will be largely fought on geography in the Third World (TW.) Driven largely by the chimera of the Gulf War, images of centers of gravity and smart weapons overshadow the tawdry back-alley settings in which actual conflict will take place in the future. In reviewing the "lessons learned" of recent wars in the Third World— where 95 percent of future conflicts will take place—a number of more accurate images appear:

> First, the mix of forces facing each other has not been very different, regardless of the type of war being fought. Navies have been conspicuous by their absence or their modest role. Air forces have played a secondary role. Almost without exception, wars in the TW have been won or lost by ground troops.
>
> Second, the type of war has not necessarily determined the tactics used; in fact, it has blurred the differences between them. Furthermore, as the frequent oscillations in tactics and strategy suggest, both irregular and regular armies in recent years have shared much in term of their ability to implement offensive doctrine. With the exceptions of campaigns involving an industrialized power, it seems that the defense has become the most effective form of warfare in the TW. With the exception of Israel and the war in Lebanon, none of the Third World combatants that have used an offensive doctrine has been able to effect a decisive breakthrough or cause the defenders to retire in disorder or retreat.
>
> Third, the duration of recent wars have generally been protracted and have engaged large arrays of national and subnational forces. Most of the wars fought between regular lesser developed country army's have not been shorter.
>
> Fourth, the involvement of external nation forces has been large. Of the wars fought between 1945 and 1976, 38 percent were fought with foreign participation.

Lastly, because guerrilla wars pose particularly difficult
challenges for governments, they represent a new and difficult
form of conflict.[14]

If the foregoing is correct, then we believe that the Infrastructure
paradigm must address the Gray Area Phenomena (GAP). GAP
is defined as threats to nation-states by non-states actors and
non-governmental processes and organizations. The Gray Areas
at once appear to be strikingly new and uncomfortably old.
Simply put, they are the most critical issues confronting the
world community as we enter into the Century. Just beyond the
horizon of current events lie two possible futures—both of which
look bleak. The first is the retribalization of large swaths of
humankind by war and bloodshed: a Lebanonization of national
states in which culture is pitted against culture, people against
people, tribe against tribe—a Jihad in the name of a hundred
narrowly conceived faiths against every kind of globalization and
interdependence.

The second is being borne by the onrush of technical,
economic, and ecological forces that demand increased
integration and uniformity. The planet, it appears, is both falling
apart and coming reluctantly together at the very same time.[15]

The GAP consists of the following "arena's of conflict":

- Ethno-religious-nationalistic conflicts;
- Weapons proliferation—both conventional and nuclear,
biological and chemical;
- Conflict over scarce resources;
- AIDS and other infectious diseases;
- The globalization of Organized Crime;
- Drug Trafficking;
- Economic Warfare and conflict over technology;
- Emigration; and,
- Famine.

The United States today is involved to some degree in each of the aforementioned arena's. While we strongly believe that traditional force-on-force warfare will always be with us, in the main, we sense that the present and future world situation more than makes the case for a strategy that encompasses the full-range of conflict supported by the appropriate technologies. Thus, we would add the Gray Area "arenas of conflict" to the DIA construct, and in so doing, move the ongoing debate to the new plateau of Information Based Warfare.

Tomorrow

> We should be seeking tentative answers to fundamental questions, rather than definitive answers to trivial ones.
>
> —James Billington

Before framing a new definition and model for Information Based Warfare, we turn to a nonwestern look at information and warfare. [It is all too soon forgotten that the works of Clausewitz are a western concept.] Shen Weiguang, a writer who appears to be at the forefront of PRC IW theorists argues that:

> in a military sense alone, information warfare refers to both sides' attempt to gain the initiative of the battle through their control over information and flow of intelligence. With the support of information, both sides intend to comprehensively apply military deception, operational secrets, psychological warfare, and electronic warfare to destroy the enemy's information systems, block the flow of the enemy's information, and create false information to affect and weaken the enemies command and control capability. At the same time, they must ensure that their command and control system is not damaged in the same way by the opponent.[16]

Further addressing the issues, Shen Weiguang illuminates a number of thoughts that cut to the heart of the concept of

46

Information Based Warfare. He correctly argues that IBW in one sense is the "quiet battlefield," something we attempted to portray in our opening scenario. Turning toward the concept of "centers of gravity" we feel that his descriptions of "attacks on the enemies cognitive and trust systems" perhaps better portrays the idea of attacking C2 and C4I nodes. This, he argues, is the main target of information warfare. The concepts of "hard attack" and "soft attack" with "soft damage" are noted as the two end-product manifestations of IBW; and, the concept of "war of structural damage" [read Infrastructure War] is thought to come into full play only when "it has absorbed the essence of information warfare."

Technology, he argues, does not determine superiority. It is determined by new tactics [in part this could be naval guerrilla warfare-PRC style] and "independent creations of commanders in the field. Information based warfare enables one to break with "traditional stylized engagement, " something we feel will give new impetus to the military arts.

Like a number of perceptive U.S. thinkers, Shen Weiguang, has seized upon the Toffler' work to understand and further flesh-out the idea of "niche-warfare." And, he appears to understand that IBW will spawn special operating forces [differentiated from special forces] in the IBW setting.

His comments about command and control are perceptive:

The operational target of information warfare lies in control rather than bloodshed;

The key to victory lies in human policy decisions rather than technology [hopefully a lesson learned and re-learned in the Naval War College games]; genuine advantage does not necessary lie in the leading technology but the leading ideas. History has given evidence to the fact when some new technology brings mankind brightness, a shadow is cast simultaneously. Advanced electronic computers and information technology link society and the army to an integrated network; the result is very high efficiency and great fragility;

The input in grasping knowledge, costs far less than directly purchasing advanced weaponry [a Sun-tzuian slip, particularly in light of the fact that the PRC has taken this approach for years as it has covertly acquired primarily mid-level technology from the West, enabling them to jump the learning curve]. [17]

A Tentative Idea

Returning again to the working definition of Information Based Warfare we suggest the following changes as noted in italics:

Information-based warfare is an approach to [*quiet war and*] armed conflict focusing on   the management and use of information in all its forms and at all levels to achieve a decisive military advantage especially in joint and combined environments [*through the   use of special operating forces that are an integral part of the overall force structure.   capable of traditional force on force war and the execution of "quiet-war*] Information based-warfare is both offensive and defensive in nature—ranging from measures that   prohibit the enemy from exploiting information to corresponding measures to assure the   integrity, availability, and interoperability elate of friendly information assets.

While ultimately military in nature IBW is also waged in political, economic. and social   arenas and is applicable over the entire national security continuum from peace to war   and from 'tooth to tail.' Finally. Information Based Warfare focuses on the command and   control needs of the commander by employing state of the art information technology   such as synthetic environments to dominate the battlefield.

48

We have added the concept of "quiet-war" because we believe that the information/communications tools that technology brings to the IBW table offers theoreticians strategists and tacticians the opportunity to plan and implement courses of action as exemplified by the opening scenario in this paper. We believe that it is this element— unattributable strikes which by their very nature preclude moving further up the conflict spectrum—is the concept that holds the most promise in the execution of Information Based Warfare.

A Tentative Model

General Capabilities	Weapons Systems
Strategic Agility	Long-range,stand-off, precision strike
Precision Intelligence	weapons[lethal and electronic]
Mission driven joint	
forces	
Information dominance	Stealthy ships and planes systems
[deception, denial, sensors, virtual quiet-war reality capable]	

Defense against chemical and	*Force Structures*
biological attack	Special operating forces w/in
overall force	structure
	Light mechanized ground forces
	Brigade sized task forces[18]

Strategy

We argue that the very soul of strategic thinking is the search for the answer to the question:

How do we win? That answer is to be found in the strategic principals:

- Ends-the protection of national interests
- Ways- the concept of how the job will get done
- Means-the resources that describe what it will take to support the concept. [19]

Answers to applying the strategic elements—in any setting in the new world disorder, with any type of force structure—are to be found in five themes suggested by Colin Gray:

First, consider strategy as a mosaic, each of the pieces of which must be understood in terms of what the sum total means.

Second, geography is the most fundamental of the factors which condition national outlooks on security problems and solutions. Geography, treated properly in political and strategic analysis is not a rigidly determining factor. The influence of geography is truly pervasive, notwithstanding the fact that influence must vary in detail as technology changes.

Third, is the use, abuse and often simply nonuse of historical experience in strategic theorizing and strategic planning. The United States in the closing years of the twentieth century is a political culture characterized by a short attention span for difficult issues of international security; by a proclivity to seek pragmatic solutions to problems which may be conditions to be accommodated rather than puzzles to be solved; and by a very noticeable historical ignorance and general disinterest.

Fourth, American strategists must begin to be aware of the influence of different national cultures upon choices for, and performance in, statecraft and strategy.

Fifth, strategists must beware of the sin of technicism—the proclivity of a materialist school of thinking about defense questions to reduce issues of means and ends to the promise in particular new military machines. Technicism shows itself in an unbalanced interest in the machine in a man-machine system (i.e., in the crossbow as contrasted with the crossbowman). But technicism refers to the disorder when that which is only technical displaces, and effectively substitutes for, that which has to be considered tactically, operationally, and strategically in far more inclusive analysis. [20]

Intelligence

Intelligence—in essence, evaluated information—is the lifeblood that courses through the strategic soul. While we could wax eloquent and long on the virtues of intelligence, it is our sense that the evaluated information that is the grist for Information Based Warfare is be found in what Sherman Kent called the "Substantive Content of Strategic Intelligence—The Speculative-Evaluative Element." This information is what the United States must know in order to be foresighted—what it must know about the future stature of other separate sovereign states [and non-state actors], the courses of action they are likely to initiate themselves [read applied alternative futures methodology], and the courses of action they are likely to take up in response to some outside stimulus. [21]

Technological Tools Driven by a "Merkava" Frame of Mind

"Merkava"(chariot of fire) was the name given to an Israeli designed, fast prototype, tank. The essence of this type of technology is that it is driven by the state of mind that recognizes dramatic, fast-moving change, and attempts to support strategy by being capable of thoughtful design and rapid fielding. In U.S. terms, it is the stuff of the "Skunkworks"—a similar belief that things can be accomplished quickly and made to work—on the battlefield.

Information Based Warfare -The New Battlefields

Exiled to France, the Ayatollah Khomeni sought to maintain contact with the faithful. A   clericwho espoused the return of his country to an Islamic theocracy—something: that many argued would move Iran back into the 16th century—the Ayatollah turned to 20th century tools to begin his crusade.

Through the medium of cassettes, he began to record his message. After being mass-produced m 20th century studios, his 16th century message was smuggled back to Iran. Those tapes

> both planted the seed and inspired the flowering of a movement that brought Khomeni back to power in 1979.
>
> —Anonymous

For at least a decade, the American military had tinkered with devices for sabotaging enemy electrical systems. Some of the results were serendipitous. In the early 1980's, during a Navy exercise code-named Hey [Rube, long strands of rope chaff—glass filaments in which metal shards were embedded—had been dropped over the Pacific Ocean as part of a standard tactic to befuddle an opponent's radar. An unexpectedly skiff wind carried some of the chaff ninety miles to the coastline, where it got draped across power lines, shorting out transformers and causing power failures in parts of San Diego. The Navy quietly settled the damages— while carefully noting the effects of its unintended attack.

> —Crusade: The Untold Story of the Persian Gulf War
> Rick Atkinson

Information warfare has as many meanings as it has proponents, detractors and observers. Airpower theorists see it on the wings of Desert Storm; tank commanders see it in the American army's Force XXI; simulator designers see it in their virtual realities...and strategic-war planners see it as a way to lay waste to whole societies. Information warfare studies are proliferating within the American armed forces almost as quickly as the networks that carry them from desktop to desktop. The confusing planoply stems from the fact that the information revolution, whether it is in uniform or mufti, relies upon the fastest technology to do the oldest things. Thus it is always a peculiar mixture of the familiar and the shockingly new. In that war is all about strategies, command and morale, it has always been about information. All war is information war, and so every aspect of fighting wars in the information age can be called "information warfare" by someone. If information driven warfare means something new, it is the use of information as a substitute for traditional ways of fighting, rather than an adjunct to them. There are three ways by which this might be achieved:

- the high technology equivalent of brute force;
- subversion; and,
- a new form of deterrence.[22]

For the purpose of this paper, we focus on the second element—subversion. It is our belief that because of the unique relationship between Americans and computers, IBW launched subversive acts—a prelude to, or the actual execution of "quiet wars"—may be accomplished. We believe that the opening scenario in this paper is both possible and plausible.

The evolution and ultimate "blurring" of military and civilian sectors came about in large part because of the growth in hardware and software. It was not so long ago, in the early 1960's, that the Pentagon provided the market for sophisticated electronics. Today, it makes up less than 1% of that market. In many ways, the military is following the revolution, not leading it. In 1962 Paul Baran of RAND developed a concept that permitted the linking of computers from distant locations. It was designed to preserve the integrity of the military command and control network in case of a nuclear attack. In 1969, the Defense Advanced Research Projects Agency (DARPA) funded the first test of the concept, and the first node was installed at RAND. The test consisted of scientists from remote locations passing findings and research notes back and forth. However, a year later it was being used like a mail box for the users and it proceeded to grow rapidly in use. In 1983, the military part and the nonmilitary part grew apart and finally split, the nonmilitary section grew up into what is now called the Internet.

For the first time in our nation's history, technology is no longer the sole domain of the military. The speed of the growth and application of technology is clearly a concern for defense planners. While the U.S. used some of the world's most advanced weaponry and technology in the Gulf War (much of which proved inaccurate) it also used a great deal of older technology, some of which dated back to the 1960's. As B.R. Inman and Daniel F. Burton Jr. note in their article "Technology and US. National Security, "the 8088 microprocessor used in the Patriot missile was developed by the Intel Corporation fifteen years ago."[23] This technology is not competitive today.

Opening and Closing Pandora's Box

In the center of Strike the battleship's commander, Captain David S. Bill, perched in his high backed padded chair. Although he occasionally glanced at the screens above, the captain's attention was largely fixed on the men clustered around four computers lining the far bulkhead. Something had gone awry with the ship's Tomahawk missile system. For reasons no one could fathom, the Tomahawk computers seemed confused, refusing to transfer the necessary commands from the engagement-planning console to the launch console. The resulting impasse—"casualty" in Navy jargon—meant the missiles could not be fired....On Wisconsin, where the scheduled launch was now just moments away, the men in Strike were running out of solutions. ..As the request for additional time flashed up the chain of command, an excited voice from one of the nearby ships crackled through Strike: "Alpha, alpha.'' This is the Paul F Foster. Happy trails.'' *Happy trails:* the code phrase for missiles away. Operation Desert Storm had begun without Wisconsin.

—Crusade: The Untold Story of the Persian Gulf

Rick Atkinson

As the most technologically vulnerable nation, the U.S. defense planners must recognize several concepts about the technical end of the IBW battlefield:

It must be assumed that everyone has access to technology, therefore the U.S. is the most vulnerable to attack;

As the U.S. ability to wage IBW increases, so to do the IBW capabilities, and response capabilities, of other nation-states and non-state actors;

While at first glance it would appear that IBW is most effective against adversaries with like technological capabilities and infrastructure—the top end of the IBW target spectrum—it must be remembered that technology comes in two forms—high and appropriate—and it is the latter that will always defeat the former. Thus, the benefits of IBW in arena's of conflict as found in the Gray Areas are a target rich environment must be pursued alongside the force on force approach.

The technical realm of information based warfare can be split into two major areas: Netwar and Cyberwar. Netwar is the information related conflict between nations, societies, governments, or non-state actors with the targets being information systems and communications, and it consists of destruction of communications, acquisition of information, release of misinformation, and destruction or deletion of data. Cyberwar consists of militaries conducting their operations according to information based theories and principles. These terms can be broken down even further. Netwar can be broken into three main categories: Personal, Corporate, and Global. We are constantly warned about the personal level of Netwar, we have passwords to computer accounts, never give our social security number over the telephone, and generally try to be careful about our own information. But protecting yourself from someone who genuinely wants to conduct information war against you is nearly impossible. Corporations are subject to a barrage of information security problems to an even greater extent. The stealing of company secrets is a new form of economic war that is plaguing many companies. It is also possible for one company to release false findings that indicate that a competitor has a poor or dangerous product. IBW spawned misinformation—a fertile area for denial and deception techniques—are very difficult to combat and correct. Finally, on the global level, Netwar can be waged against industries, economies, nations, or non-state groups. It ranges from intelligence and information leads, to the terrifying possibilities of terrorist groups breaking into such communications as an air traffic control system at a major U.S. airport. The impact at the global level can be much more wide spread and all threatening than that of the personal or corporate levels of netwar.

On both sides of the war, Netwar and Cyberwar, the key to good protection and offense is good software to run the systems, encode the data, and peck holes in other's data. The vital link for the information and the technology is software, which relies heavily upon implanted logic. The newest and brightest theories and concepts of logic center around what is called 'fuzzy logic'. The name fuzzy logic makes one think of a system that is inexact, but that is not the case, instead it has an infinite exactness. Fuzzy logic expands our normal computer principle of 0 or 1, true or false, and in the place of this is infinite degrees between 0 and 1. Fuzzy logic is used every day in the development of everything from smart washers to train systems in Japan. With a smart washer, you merely press a start button and the machine figures the cycle, how much detergent, and the water use; the logic train systems are smooth as silk, quick, and always on time. The theory started with a few paradoxes one of which is The Paradox of Theseus' ship:

> When Theseus returned from slaying the Minotaur, says Plutarch, the Athenians preserved his ship, and as planks rotted, replaced them with new ones. When the first plank was replaced, everyone agreed it was still the same ship. Adding a second plank made no difference either. At some point, the

Athenians may have replaced every plank in the ship. Was it a different ship? At what point did it become one?

At different points in time the ship was certain degrees the original ship, for instance when half the planks were replaced, it was .5 the original ship. This same principle of degrees can be applied to many concepts around us, and what you end up with is a very exact and reliable process for computers. When you tie this concept into ideas such as artificial intelligence, you may obtain a computer that has the capacity to learn because it can start to work with more than 0 and 1.

Because of the importance of the technical tools of IBW—and their fragility—we list below the possible contents of an IBW techno "kit-bag":

Computer Viruses - A piece of code, or code fragment, designed to begin replicating itself when a host program begins to run. It's objective is to erase data, software programs, or memory in order to interrupt the action of the computer it infests. These can be loaded unwittingly by a user off the Internet or some other disk.

Worms - Instead of being just a code fragment, a worm is an entire program in and of itself. The code begins to replicate as soon as it touches a computer system, not dependent on the start of execution of any other program. Its aim is to eat up computer resources and/or delete data to result in a crippling effect on the host computer.[We argue that it is such worms could cause the damage we allude to in the opening scenario.]

Trojan Horses - As the name would suggest, a Trojan horse is a program, or code fragment inside a program, that performs a function unbeknownst to the user. It usually performs a simple task on the outside, while unleashing a virus or a worm on the inside. It can also perform information retrieval while performing a task, leaving no trace.

Logic Bombs - Similar to a Trojan horse, a logic bomb is used to release a virus, worm, or complete some other secret task. These are usually planted by a programmer or system developer. This could be useful in U.S. military strategy, instead of open bombs, the U.S. could plant logic bombs to retrieve specific data from foreign users. This is plausible because many of the main software development in the world occurs in the US.

Trap Doors - The use of this is similar to a logic bomb. It is a secret way back into a system left by a programmer or designer. Unlike a logic bomb that executes, a trap door is merely an unknown security flaw.

Chipping - This is the term for doing any of the above to the hardware of a system instead of to the software. One can build in circuitry to a chip that performs specific functions.

Machines and Microbes - Although one thinks of Star Trek and other science fiction when the subject of nanny machines comes up, it is actually a feasible plan. machines and microbes are tiny machines, like a small insect, that "eat" electronic circuits, oil, plastics, etc. If unleashed on a computer center, the result would be total shut down.

Electronic Jamming, HERF guns, and EM bombs- These are all forms of electronic signal jamming. One can jam communications, stopping computer information flow; shoot radio signals at an electronic target to shut it down with a High Energy Radio Frequency (HERF) gun; or send out a high powered electromagnetic pulse, an EMP bomb. [In the Gulf War, the allies used 35,000 different radio bands.]

Hitch-hiking- This is a group of people, a subsection of hackers, that merely sit on the net and wait. They wait for any information and interesting code to go by and then hitch a ride with it, they follow and delete it, or they simply copy and steal it.

Thinking in terms of closing Pandora's Box, the following techniques will have to be considered as an IBW operational security plan is built:

Firewalls - A limited gateway to the Interment from a company or group. Only passwords and certain configurations can get in, and everything is checked for viruses, etc. Only privileged personnel, or personnel with certain system configurations can get out. If implemented correctly they can be useful.

Encryption - This growing area of interest consists on coding data so that others can not place bombs in or around it or read the data. This is growing more difficult as hackers and other groups become more efficient at breaking the codes. This is one of the areas that fuzzy logic could come into serious play.[24]

If, at first glance, the foregoing appears to be merely a "laundry list," a reading of the following paragraph will surely prove sobering to the IBW planner:

The Defense Information Security Agency (DISA) conducts vulnerability studies of military and government computer systems. Their figures are truly alarming: 88% of defense computer systems are easily penetrated. Of the successful penetrations, 96% are not detected. Even worse, 95% of the detected penetrations are not reported or responded to. Even when an intrusion is detected, it is usually impossible to determine who did it. DISA studies indicate that there were possibly 300,000 intrusions into government computer systems in 1994 alone.[25]

A Distant Bugle

Despite incessant barbarian incursions and major military threats throughout its history, Imperial China was little inclined to pursue military solution to aggression—except during the ill-fated expanionistic policies of the Former Han dynasty, or under dynamic young rulers, such as T'ang Tai-tsung, during the founding years of a dynasty. Rulers and ministers preferred to believe in the myth of cultural attraction whereby their vastly superior Chinese civilization, founded upon Virtue and reinforced by opulent material achievements, would simply overwhelm the hostile tendencies of the uncultured. [26]

Today, there appears to be a striking similarity between a number of western nation-states and ancient China. This likeness is a reflection of a malady that continues to run its course through history as kingdoms, empires, and nation-states have lost sight of the fact that survival is always won—and maintained—at the point of the sword.

Modern states, especially those who have triumphed in the Cold War and have the greatest interest in preserving peace, and most particularly the United States, on whom the burden of keeping the peace must fall, now, and in the foreseeable future, are quite different. The martial values and the respect for power have not entirely disappeared, but they have been overlaid by other ideas and values, some of them unknown to the classical republics. The most important of these is the Judaeo-Christian tradition, and especially the pacifist strain of Christianity that emphasizes the sermon on the mount rather than the more militant strain that played so large a role over the centuries. Even as the power and influence of formal organized religion have waned in the last century, the influence among important segments of the population of the rejection of power, the evil of pursuing self-interest, the wickedness of war, whatever its cause or goal, have grown. There are now barriers of conscience in the way of acquiring and maintaining power that would have been incomprehensible to the Greeks and Romans. In spite of their victories in the Cold War and, more recently, in the Gulf War, the United States and its allies, the states with greatest interest in peace and the greatest power to preserve it, appear to be faltering in their willingness to pay the price in money and the risk of lives. [27]

What seems to work best, even though imperfectly, is the possession by those states who wish to preserve the peace of the preponderant power and of the will to accept the burdens and responsibilities required to achieve that purpose. They must understand that no international situation is permanent, that part of their responsibility is to accept and sometimes even assist changes, some of which they will not like, guiding their achievement through peaceful channels, but always prepared to resist, with force if necessary, changes made by threats or violence that threaten the general peace. [28]

Information Based Warfare gives the sword another edge.

Notes

1. Foreign Broadcast Information Service, 13 Dec 95, p. 25 (hereafter FBIS).

2. FBIS, p. 23.

3. FBIS, p. 24.

4. *The Seven Military Classics of Ancient China,* p. 277 (hereafter Classics).

5. FBIS, p. 25.

6. *Joint Force Quarterly,* Autumn/Winter 1994-95, p. 29.

7. Earl H. Tilford, Jr., *The Revolution in Military Affairs: Prospects and Cautions* p. 3 (hereafter Tilford).

8. Tilford, p. vi.

9. Steven Metz and James Kievit, *The Revolution in Military Affairs and Conflict Short of War*, p. vi.-vii (hereafter Metz & Kievit).

10. Metz & Kievit, p. vii.

11. Metz & Kievit, p. vii.

12. Metz & Kievit, p. 3.

13. Intelligence Support to Infrastructure Warfare, Directorate for Combat Support National Military Intelligence Production Center, Defense Intelligence Agency p.2-3. (We are grateful to Louis Andre for allowing us to reference and respond to his excellent White Paper.)

14. Robert E. Harkavy and Stephanie G. Neuman, *The Lessons of Recent Wars in the Third World: Approaches and Case Studies*, Vol. 1, p. 283-284.

15. J.F. Holden-Rhodes and Peter A. Lupsha, *The New Horsemen of the Apocalypse*, p.1.

16. FBIS, p. 24.

17. FBIS, pp. 22-29.

18. Michael J. Mazarr, *The Revolution in Military Affairs: A Framework for Defense Planning*, p. 33. (We seized upon Mazarr's model as the closest we could find to our thinking and modified it to reflect the thrust of our argument.)

19. Drawn from: *Campaign Planning, Strategic Studies Institute. U.S. Army War College*

20. Colin S. Gray, *War, Peace, and Victory: Strategy and Statecraft for the Next Century*, p. 15-16.

21. Sherman Kent, *Strategic Intelligence for American World Policy*. p. 39-68.

22. "Information Advantage," *The Economist*, 10 June 95, p. 38.

23. B.R. Inman and Daniel F. Burton Jr., *Technology and National Security*

24. Reto E. Haeni, *An introduction to Information Warfare.*

25. Commander William E. Rhode, U.S. Navy, "What is Information Warfare?" *Naval Institute Proceedings*, p. 37-38.

26. Classics, p. 2.

27. Donald Kaman, *On the Origins of War*, p. 570-571. (hereafter Kaman)

28. Kaman, p. 571.

⭕ INFORMATION TERRORISM:
Can You Trust Your Toaster?

Matthew G. Devost, Brian K. Houghton,
and Neal A. Pollard
Science Applications International Corporation

Scenario: September 1998

Tensions in the Balkan conflict have grown geometrically, particularly through Croat and Muslim aggression, with the failure of a series of peace accords. A new peace accord has been worked out, brokered by the United States, that stands a chance to redeem U.S. and NATO policy failures in the region, although some see it as harsher on Serbian combatants while it acquiesces to Croatian demands. Furthermore, NATO efforts at economic reconstruction have been particularly biased against Serbian interests. Determined to see its success in the face of flagging Congressional and public support for prolonging Bosnian operations, the President has increased the U.S. military presence in the region, establishing a new NATO airfield in Brcko, on the Bosnian/Croatian border, to facilitate logistics and put an end to the Balkan conflict. In September, with the prolonged fighting and the oncoming winter and its attendant fuel and food shortages and wave of refugees, stability in the region begins to deteriorate and Croat and Muslim troops increase activity; the President increases airlifts of troops and materiel, to counter tensions and support peace initiatives.

During the successive peace accord failures, and in response to increasing Croatian and Muslim aggression, sluggish economic recovery, and a tendency for NATO to be biased against Serbs, a group called the Serbian Council for the Liberation of Bosnia (SCLiB) is formed, consisting of Serb paramilitaries in Bosnia, Yugoslavia, and abroad, who have political and military influence among Yugoslavian and Bosnian Serb officials; the Council also consists of students in Slovenia, Hungary, and Yugoslavia, many of whom lost family members at the hands of Croats or NATO troops. The Council coalesced once members began to meet and communicate via the Internet, using PGP encryption to hide their interests and intentions. Their primary objective is revenge, to redress grievances from Croatian land usurpation and its support by their American patrons, and to rid the area of the NATO presence by dramatizing their cause to the people of the world, influencing them, and thus their governments, to demand NATO leave the area.

Having garnered enough financial and operational support through usual terrorist means, the Council formulates an attack, beginning with the CNN Web Page. By accessing the CNN Weather forecast, the Council times their attack for a night of intense storms in the Brcko area. Paramilitary members of the Council intrude on the frequencies of the approach and tower radios at the Brcko airfield: an airfield recently set up, and thus lacking ideal security measures, procedural experience, and full integration of NATO countries' respective military communications systems. In the storm, flying into the airfield with its navigation lights off due to reported ground fire, a full C-130 troop transport is cleared to land by the approach intrusion. Another C-130, laden with fuel and also with its lights off, is cleared for take-off on the active runway, by the tower intrusion. The landing C-130 crashes into the second C-130. The resulting crash kills all aboard both planes. After hearing the explosion from their vantage point on a nearby hill, the intruders send a cellular signal to awaiting Council hackers in Slovenia. Upon receipt of the signal, the hackers immediately issue an "e-communiqué," taking responsibility for the crash, explaining how it was done, and giving the location of the intrusion equipment used, on which is engraved "SCLiB." The remainder of the message is their manifesto and claim for redress of grievances against life, property, and national identity. The end of the message is an invitation and address to access their Web site, which is actually run from a computer in Amsterdam by Slovenian foreign exchange students, via an anonymous web service account in Finland. This message is sent to and received by every major print and electronic news organization in the industrialized world, before the debris from the C-130 crash had settled.

The resultant publicity is astounding: CNN, Reuters, ITAR-TASS, and AP immediately broadcast the message, with the Web address. In addition, the e-communiqué itself was sent out to over 30,000 e-mail addresses in the first hour after the crash. Six minutes after the e-communiqué had been received, the Council Web page received its first hit.

Twenty-four hours after the C-130 crash, the Council Web had received over 1 million hits. The Web page was dramatic and rife with propaganda and claims against American, NATO, and Croatian imperialism and atrocities in the Balkan region, and included questionable allegations of illegal arms transfers between NATO governments and Bosnian Muslims and Croats. Several references were included to the former U.S. presence in Lebanon, and how that presence was resolved. Twenty-four hours after the first hit, the first accessing system crashed, with all files irretrievably deleted, as a result of a Trojan horse the Council hackers had embedded in the Web page, exploiting a flaw in the programming language similar to one discovered by Princeton computer scientists in February 1996.[1] The flaw allowed a webmaster access to the hard drive and files of the machine that had unwittingly accessed the tainted Web page. Exploiting this flaw, the Council embedded a program that

65

activated 24 hours (according to the system internal clock or any other time-keeping mechanism the machine could access) after the page was hit, destroying the functions and files of the system it infected. Although this created a sensational climate of fear throughout the computerized civilian world, the most damage done was to investigative and defense organizations, who immediately and naturally accessed the Web page before most of the news organizations had disseminated its address. This included the American Department of Defense, the Defense Ministries of all NATO countries, the American Department of Justice and Treasury, and the Central Intelligence Agency. Final damage to unclassified systems was incalculable, but the dramatization of the Council's cause was greatly effective. Since the Trojan horse was set to activate 24 hours after the Web site had been hit, computer failure rates tended to cascade, and were slow in tapering off, despite warnings to avoid the terrorists' Web page.

The actual reports of the carnage of the crash reached the public: these reports, on top of the fear created by the computer disasters, and the general frustration with American efforts in the Balkans, put enormous pressure on Congress and the President. Because of a lack of treaty conventions, American investigative agencies were not allowed to violate protocols of Finland's cyber-community; thus, investigators were unable to ascertain the identity of the anonymous server's customer, or the location of the Web site in Amsterdam. The Council's information terrorists remained secure in anonymity, and their success in hiding prompted many copy-cat web pages, a spate of "Internet liberators," and re-circulation of the Council's original manifesto and web page detail. With Congressional elections just over a month away, the Balkan mess became a rallying point of congressmen to pressure the President. Finally, the President had little choice but to accede to the public's and Congressional demands to bring the boys back home. Without American logistical and operational support, NATO's presence and power in the region was significantly reduced.

As with most conventional terrorist attacks, tactical damage to military and government information systems was relatively small (although several billion dollars of civilian and commercial information value could conceivably be lost in such a web-based attack). However, the strategic objective was not damage: as with most conventional terrorist attacks, the strategic objective was publicity, drama, and leverage to influence public and policy. The terrorists achieved their strategic objectives, clearly and effectively. [2]

Introduction

In the remainder of the paper the authors will: 1) define information terrorism within the context of information warfare[3] as well as conventional terrorism; 2) offer a possible response to the phenomenon of information terrorism.

Information and Stability: The Lure of Technology

Extremist groups often resort to political violence when they lack the power to achieve political objectives through non-violent legal means. In an effort to attract the attention of the public, political terrorists perpetrate their acts with the media at the forefront of their strategy: this strategy calculus is based on the assumption that access to the communication structure is directly related to power.[4] Believers in this assumption might target digital information systems in pursuit of political goals.

The National Information Infrastructure (NII), and Global Information Infrastructure (GII) support financial, commercial and military information transfers for consumers, businesses, and countries. Considering the presence of computers in modern society, it is not surprising that terrorists have occasionally targeted computer systems in the past. A "PLO" virus was developed at Hebrew University in Israel; in Japan, groups have attacked the computerized control systems for commuter trains, paralyzing major cities for hours; the Italian Red Brigade's manifesto specified the destruction of computer systems and installations as an objective for "striking at the heart of the state."[5] More recently, Sinn Fein supporters working out of the University of Texas, Austin, posted sensitive details about British army intelligence installations, military bases, and police stations in Northern Ireland on the Internet.[6] Terrorism is a rapidly evolving and responsive phenomenon. Terrorist technology and tactics are sensitive to their target political cultures, and have progressed at a rate commensurate with dominant military, commercial, and social technologies.

As technology becomes more cost-effective to terrorists--that is, its availability and potential for disruptive effects rise while its financial and other costs go down--terrorists may become more technologically oriented in tactics and strategies. In 1977, terrorist expert Robert Kupperman, then Chief Scientist of the U.S. Arms Control and Disarmament Agency, recognized that increasing societal reliance upon technology changes the nature of the threat posed by terrorists:

> Commercial aircraft, natural gas pipelines, the electric power grid, offshore oil rigs, and computers storing government and corporate records are examples of sabotage-prone targets whose destruction would have derivative effects of far higher intensity than their primary losses would suggest....Thirty years ago terrorists could not have obtained extraordinary leverage. Today, however, the foci of communications, production, and distribution are relatively small in number and highly vulnerable.[7]

The incorporation of information technology in the military-industrial complex, and the design and implementation of information warfare strategies, may also draw terrorists to computer technology. In the final days of the Cold War, NATO allies took seriously the premise that as warfare grows more electronic and dependent upon information technology, the vulnerabilities and risks of sabotage grow.[8] In a RAND paper,

Dr. Bruce Hoffman asserts that, because of the operational conservatism resulting from the terrorists' "organizational imperative to succeed":

> ...terrorists will always seek to remain just ahead of the counter-terrorism technology curve: sufficiently adaptive to thwart or overcome the countermeasures placed in their path but commensurately modest in their goals (i.e., amount of death and destruction inflicted) to ensure an operation's success.
>
> In this respect, rather than attacking a particularly well-protected target-set or attempting high risk/potentially high payoff operations, terrorists will merely search out and exploit hitherto unidentified vulnerabilities and simply adjust their plan of attack and tactical preferences accordingly.[9]

Information technology offers new opportunities to terrorists with the above strategic concerns. In pursuing this modus operandi, a terrorist organization can reap low-risk, highly visible payoffs by attacking information systems.

Defining Information Terrorism

Information warfare has been examined within the context of state-on-state operations, as well as assessments of peer or near-peer competitors. However, sub-state and gray area[10] phenomena, especially information terrorism, have yet to be addressed within the paradigm of information warfare. Information warfare emanating from the low intensity end of the political violence spectrum represents a threat to American national security and defense.

An act of political violence by anyone other than a member of the armed forces of a legitimate state is often branded an act of terrorism. This is only occasionally correct[11], but the criminal and subversive connotations of the term "terrorist" have resulted in many acts of computer abuse being labeled "information terrorism." These acts have ranged from using personal information for extortion, to hacking into a network, to physical and/or electronic destruction of a digital information system. This is too simplistic a taxonomy for such a complex phenomenon.

Labeling every malicious use of a computer system "terrorism" serves only to exacerbate confusion and even panic among users and the general public, and frequently hinders prosecution and prevention by blurring the motivations behind the crime. Furthermore, political crimes have vastly different implications for national security and defense policy, than other "common" crimes. Terrorism is a *political* crime: an attack on the legitimacy of a specific government, ideology, or policy. Hacking into a system to erase files out of sheer ego, or stealing information with the sole intent to blackmail, is nothing more than simple theft, fraud, or extortion, and certainly is not an attack upon the general legitimacy of the government. Policy and methodology to counter crime depends a great deal upon criminal motivations;[12] thus, clearer and more concise definitions of "information terrorism" are needed, if it is to be addressed by national security policy. Attacks on the legitimacy of a government or its policies are not "common" criminal motivations. The quasi-criminal, quasi-military nature of terrorism blurs the distinction between crime and warfare. Distinctions between law enforcement and military duties become equally blurred,[13] and can be clarified only through coherent policy dictating those duties, based upon a clear view of the nature of the enemy.

Political terrorism is the systematic use of actual or threatened physical violence in the pursuit of a political objective, to create a general climate of public fear and destabilize society, and thus influence a population or government policy. Information terrorism is the nexus between criminal information system fraud or abuse, and the physical violence of terrorism. However, particularly in a legal sense, information terrorism can be the intentional abuse of a digital information system, network, or component toward an end that supports or facilitates a terrorist campaign or action. In this case, the system abuse would not necessarily result in direct violence against humans, although it may still incite fear. Most terrorism scholars, when defining "political terrorism," would include physical violence as a necessary component; thus, many acts of criminal computer abuse would not be considered terroristic, if they do not result in direct physical violence. However, scholars must face the fact that as technology's implications broaden on society and politics, social and political definitions should likewise broaden to accommodate technology.[14] The semantic vacuum of a universally accepted comprehensive definition leaves room for considering information system abuse as a possible new facet of terrorist activity.

Tools and Targets

In a Third-Wave[15] society, there are two general methods in which a terrorist might employ an information terrorist attack: (1) when information technology is a target, and/or (2) when IT is the tool of a larger operation. The first method implies a terrorist would target an information system for sabotage, either electronic or physical, thus destroying or disrupting the information system itself and any information infrastructure (e.g., power, communications, etc.) dependent upon the targeted technology. The second method implies a terrorist would manipulate and exploit an information system, altering or stealing data, or forcing the system to perform a function for which it was not meant (such as spoofing air traffic control, as highlighted in the third scenario).

In the matrix below, cell (a) addresses "traditional" terrorism (e.g. hijacking, bombings, assassinations, hostage taking, etc.) The authors consider cells (b), (c), and (d) to be information terrorism. Cell (b) represents a low tech solution for a high tech target (e.g. the IRA attack on Square Mile financial district of London[16]). Cell (c) exploits information systems to wreak physical damage. Cell (d), digital tools against digital targets, exploits vulnerabilities in military, commercial and civilian/utility systems that rely on information technology. The authors believe cell (d) to be "pure" information terrorism and likely the most difficult to detect and counter.

Tool		Target	
		Physical	Digital
Tool	Physical	(a) Conventional Terrorism (Oklahoma City Bombing).	(b) IRA attack on London Square Mile, 4 October 1992.
	Digital	(c) Scenario (Radio intrusion in C-130 crash).	(d) Trojan horse in public switched network.

Figure 1

No Symmetrical Response

A dilemma of combating terrorism in a democratic society is finding the right balance between civil liberties and civil security. Military operations within a democratic society, even to "protect" it, often are inconsistent with the principles of that society. The military thus confronts a paradox as it strives to combat terrorism. Although terrorists can use brutal, indiscriminate force against the military and civilian population, the military response may be limited. If the perpetrator of a terrorist action is found to be state-sponsored, a military response against state targets is possible (e.g. United States sending F-111s against Libya in response to Berlin Disco bombing in 1986).

Frequently terrorists are not state-sponsored, but are hidden within the civilian population. Tanks, aircraft and cruise missiles are ineffective against an enemy that blends itself into a civilian background. Information terrorists, outside the United States[17] have an easier means of disappearing inside their civilian population. Operating from homes[18] via modems, these terrorists

72

can functions in their cell like structure using encrypted e-mail as means of communication to their organization's network, and thereby reducing their chances of exposure.

The U.S. government faces this same paradox as it confronts information terrorism. Military, civilian and commercial databases, computer systems, information infrastructures all are potential targets of information terrorists. Whether through digital or physical means, the information terrorists can destroy, disrupt, degrade, deny or delay vital information that the military relies upon, and thus become a threat in peace time, as well as in time of war. How can the U.S. national security establishment respond to the informational attacks of terrorists, when the terrorists hide behind a veil of digital anonymity? How much of information terrorism is a military concern and how much is within the jurisdiction of federal law enforcement?

The U.S. military could find it difficult to respond against a small and digitally networked enemy such as a terrorist campaign. The U.S. national security establishment needs to use a flexible, integrated response to counter information terrorists— one which employs information warfare tactics tailored to counter gray-area phenomena, but also reserves the use of conventional counterterrorism operations.

Recommendations

The U.S. national security establishment must be equipped to respond militarily to information terrorism. Firstly, the military will always be a target of terrorism. Furthermore, the information terrorism attack may be state-sponsored and the first wave of a "digital Pearl Harbor." Origins of digital attacks are usually difficult to discover at first, and if the attack is indeed a precursor of peer or near-peer information warfare, a military response will be required.

However, democratic societies must carefully weigh the use of military forces in the prevention and countering of terrorism, even though their militaries may be targets of the attacks. By calling in the military to respond to conventional terrorist actions, the terrorists and their cause may achieve a degree of legitimacy. The terrorists actions then have escalated from a criminal level to a "enemy of the state." This quandary can be avoided when countering information terrorists. There are no visible soldiers on the streets to heighten civilian anxieties when using digital attacks to counter the terrorists. The military's response, like that of the information terrorists, can be anonymous, fully networked, and swift.

The military has unique capabilities to confront and counter international information terrorism which the domestic law enforcement agencies lack, particularly in the military's specialized training and established international presence. Aspects of an international information terrorist attack (especially within cell (d) [see Figure 1]) would fall squarely within the jurisdictions of several federal law enforcement agencies because these attacks would affect a domestic information system, just by virtue of the connectivity of such systems. Furthermore, the investigative abilities of law enforcement agencies such as the FBI and the Treasury Department's FinCEN (Financial Crimes Enforcement Network) are particularly well-suited to counter information terrorism, from detecting the logistics and method of attack to following the money trail and uncovering a possible sponsor. The most important aspect of any counter terrorist endeavor is a rapid response time. Law enforcement is particularly adept at rapid crisis management. Clearly, the ideal response structure would be one that incorporates assets from both the military and law enforcement. Such a structure could also incorporate the military in an advisory role in domestic incidents, and likewise, law enforcement assets in an advisory role in overseas incidents.

Offensive information warfare techniques developed for military use at a state level could also be utilized to respond to information terrorism. Law enforcement agencies, in general, do not have similar offensive information warfare capabilities. For this reason a specialized and integrated counter information terrorism group is required. These highly trained information warriors would be the national security equivalent of Carnegie Mellon's Computer Emergency Response Team, but with an offensive capability. Like a "Digital Delta Force" these Digital Integrated Response Teams (DIRTs) would work from remote computer systems and use information warfare tactics to detect, locate and counter the information terrorists. The DIRTs would be in networked remote cells inside CONUS (with one on the East and West coasts, and an additional cell in the Midwest). The DIRTs would exploit law enforcement IT-oriented assets, investigative capabilities, and intelligence bases. The DIRTs, created by Executive Order, would operate as a cell of the National Security Council and take its directives from the information terrorism counterpart to the White House "Drug Czar."

These information warriors, comprised of members from the Joint Services, as well as Justice and Treasury Departments, would strike using digital means against computers and networks used by the information terrorists. Using an anonymous response, the U.S. government could strike at information terrorists without large display or legitimizing the terrorists, both of which would occur with a physical response. Such a response offers ultimate plausible denial. In addition, the DIRTs close integration with law enforcement agencies would provide legal guidance and accountability, and avoid a "Posse Comitatus" syndrome.

This structure would combine the investigative and jurisdictional assets of the law enforcement community with the offensive capabilities of the military. If the United States is going to enter the Information Age, we need to have policy that spans the spectrum of information-related threats to our national security, driving offensive *and* defensive assets that can respond symmetrically and effectively. Our offensive capabilities against peer or near-peer competitors are formidable, whether in information or conventional warfare. However, the integration of law enforcement assets is necessary to respond effectively to a networked gray-area attack. Without an integrated, fully articulated response policy, information terrorists could severely damage the infrastructures of our military or society, in the time it takes to argue about whose job it is to respond.

Notes

1. See http://www.cs.princeton.edu/~ddean/java/dns-scenario. html for the DNS attack scenario that Princeton researchers used to exploit a flaw in Java.

2. By a most unfortunate coincidence, this scenario was fully developed four days before the tragic crash of Commerce Secretary Brown's airplane in Croatia. While not wishing to exploit such a tragic loss, we feel the scenario is still clearly relevant. Our most sincere condolences go to the families, friends, and colleagues of all who perished.

3. For the purposes of this paper, "Information warfare" will be defined as offered by the Department of Defense: "Actions taken to preserve the integrity of one's own information system from exploitation, corruption, or destruction, while at the same time exploiting, corrupting, or destroying an adversary's information system and in the process achieving an information advantage in the application of force." (Proposed: JCS Pub 1-02).

4. See Alex P. Schmid & J.F.A. DeGraaf, *Violence as Communication: Insurgent Terrorism and the Western News Media*. Beverly Hills, CA: Sage, 1982.

5. Philip Fites, Peter Johnson, & Martin Kratz, *The Computer Virus Crisis,* Second Edition. New York: Van Nostrand Reinhold, 1992 (p.63).

6. London Times, via CNN Web News Digest, 26 March 1996 (http://www.cnn.com).

7. Robert Kupperman, "Facing Tomorrow's Terrorist Incident Today." Washington, DC: U.S. Department of Justice, Law Enforcement Assistance Administration, 1977. Cited in Grant Wardlaw, *Political Terrorism*, Second Edition. Cambridge: Cambridge University Press, 1989 (p.26).

8. Gerald Segal, "Asians in Cyberia," *The Washington Quarterly*, v.18 n.3 (Summer 1995), pp.12-13.

9. Bruce Hoffman, "Responding to Terrorism Across the Technological Spectrum," RAND Corporation, April 1994 (pp.29-30).

10. "Gray-area phenomena" is political violence that is not easily seen to be sponsored by or connected to a state or an established organization.

11. For example, there is a distinct difference between terrorism and guerrilla warfare (See Walter Laqueur, *The Age of Terrorism*. Boston: Little, Brown & Co., 1987 [p.5]).

12. This assumption is based on the notion that in political crimes, as opposed to crimes for ego or greed, the perpetrator and the beneficiary are usually not the same person, with more tenuous connections, and the intended long-term gains from the crime are usually abstract. From a law enforcement perspective, this places primacy on the motivations of the act; however, from a military perspective, it is the act itself which merits focus, since it is the act itself which wreaks the damage and poses the threat to national security. This is one of many facets in the argument surrounding the degree to which counterterrorism should be approached from a military vis-à-vis law enforcement perspective. This argument further emphasizes the need for a symmetrical, part-military part-law-enforcement response.

13. Review of Richard Hundley and Robert Anderson, "Security in Cyberspace: An Emerging Challenge for Society," from That Wild, Wild Cyberspace Frontier. Internet source: http://www.rand.org/publications/ RRR/RRR.fall95.cyber/wild.html, 5 April 1996.

14. Stephen Sloan, "Terrorism: How Vulnerable is the United States?" Internet Source: The Counter-Terrorism Page, http://www.terrorism.com/ Pubs/sloan.htm.

15. Physical violence from terrorism uses Toffler's Second Wave technology, whereas information attacks would fall within the Third Wave paradigm. (See Alvin Toffler, *The Third Wave*. New York: Willam Morrow & Co., Inc., 1980.)

16. The IRA were specifically targeting the Square Mile of London on a weekend to minimize casualties but maximize damage to a financial center of Western Europe.

17. The authors chose to not to discuss domestic information terrorism, since that falls under the jurisdiction of the FBI, but much of

the debate is similar regarding FBI capabilities to counter this threat.

18. Traditional" terrorists generally operate in an urban environment often without an established geographical locus. Information terrorism further diminishes geographical constraints through the nature of digital connectivity.

⭘ INFORMATION WARFARE:
The Organizational Dimension

Colonel Brian Fredericks
U.S. Army

Introduction

Today with all of the various interpretations and multiple definitions, Information Warfare (IW) remains an enigma. Since the Department of Defense formally published the original classified directive on IW in December 1992, the Services, Office of the Secretary of Defense, and a wide range of joint activities have expended considerable effort examining this issue. While there is a general recognition that IW has the potential to serve as an important force multiplier, the concept remains in its infancy. Ultimately the success of IW as a decisive component of U.S. national security in the 21st century depends upon achieving a viable IW architecture. This architecture must comprise three key areas: policy/doctrine, organization/training, and requirements/technology. Much has been written, discussed, and even debated on the need for overarching national policy in this area, as well the multitude of capabilities and vulnerabilities stemming from our increased reliance on advanced technology. However, a similar focus on the organizational component of IW has not occurred.

This paper specifically addresses the role of organizations as an essential element in developing and implementing a viable IW strategy. To provide a common reference point, the paper begins by defining IW. Next it analyzes the progress achieved to date in institutionalizing IW by assigning responsibility to specific organizations. Both the progress achieved within DOD and the significant challenges remaining to be overcome at the interagency level are examined. The paper concludes with a set of recommendations on how to better organize the IW effort and enable it to emerge as a decisive element of U.S. national security strategy in the 21st century.

IW Defined

Information Warfare (IW) was formally launched in December 1992 with the dissemination of DOD Directive 3600.1.[1] From the outset, widespread discussion and understanding of IW were hampered by its Top Secret classification.[2] In September 1995 the Assistant Secretary of Defense for Command, Control, Communications and Intelligence (ASD(C3I)) published the formal DOD unclassified definition of IW:

> Actions taken to achieve information superiority by affecting adversary information, information-based processes, and information systems, while defending one' s own information, information-based processes, and information systems.

This description clearly underscores both the defensive, as well as offensive aspects of IW. In the summer of 1994, the Defense Science Board[3] (DSB), drawing heavily from expertise within DOD, published the most comprehensive and authoritative discussion of IW to date. The report highlighted the distinction between information *in* warfare and information warfare. Information in warfare pertains to "getting [information] it where it is needed in a timely and reliable manner.[4]" It encompasses the collection, processing, and dissemination of information and is synonymous with the "C4I for the Warrior" vision released by the Joint Staff in 1992. C4I for the Warrior addresses the concept of a global Command, Control, Communications, Computer, and Intelligence system directly linking military units around the globe in an interoperable, fully integrated fashion spanning the range of military operations from peace to war.[5] Information in warfare capitalizes on the national information infrastructure (NII). Characterized as an "information highway,"[6] the NII is the growing worldwide information infrastructure which transcends industry, media, and the military and includes government and non-government entities. Most activities now rely on the information infrastructure including the banking' transportation, manufacturing, and electrical power industries. The Defense Information Infrastructure (DII) is an integral part of the NII with over 95 percent of DOD' s worldwide telecommunications needs satisfied by commercial telecommunications carriers.[7] Military activities relying on the DII include transportation, logistics, financial, manpower, and personnel and training.

While C4I for the Warrior focuses on harnessing ever-increasing computer storage and exchange capabilities, IW targets these information systems. The distinction between C4I for the Warrior and IW is extremely important. IW employs offensive techniques such as deception, electronic jammers, munitions and advanced technologies to deceive, deny, exploit, damage, or destroy adversary information systems, while at the same time protecting friendly information systems from disruption, exploitation and damage by an adversary.[8] The target of IW may range from influencing national level decisionmakers to corrupting the automated control of transportation systems.[9] defensive IW protects friendly information systems from disruption, exploitation and damage by an adversary.[10]or example, the Army's ongoing digitization of the battlefield is an application of information in warfare at the operational and tactical levels. Defensive IW, on the other hand, focuses on identifying and protecting vulnerabilities which arise from this increased reliance on technology.

The emergence of Command and Control Warfare (C2W) has been fundamental in understanding IW.[11] In article in *Signal* magazine LTGen James Clapper, then Director Defense Intelligence Agency (DIA), wrote "the closest description of information warfare might be found in the definition of command and control warfare."[12] IW and C2W, however, are not interchangeable terms. C2W is a subset of IW. The Chairman of the Joint Chiefs of Staff (CJCS) Memorandum of Policy (MOP) 3O, states "C2W is the military strategy which implements IW on the battlefield."[13] C2W is designed as an essential part of an overall theater campaign plan. It is implemented during "joint military operations when U.S. military forces unilaterally or as part of an allied/coalition force are opposed or threatened by an organized military or paramilitary force."[14] C2W focuses on an adversary's military command and control when military force is applied. On the other hand, the use of the word "warfare" in the term IW does not limit IW to a military conflict, declared or otherwise. is IW targets the entire information infrastructure of an adversary - political, economic, and military throughout the continuum of operations from peace to war.

Organizational Imperative

IW, as this definitional discussion highlights, is a complex issue. Organizations are essential in actually implementing this new concept and achieving a viable IW architecture. IW will only become institutionalized if activities actually take responsibility for planning and executing IW. Today IW offices have been stood up throughout DOD focusing on offensive and defensive IW capabilities, but for the most part the budgets and staffs of these elements are very limited. They represent an important start in what will likely be a long and slow process.

IW Executive Board: The Deputy Secretary of Defense chairs an IW Executive Board established in May 1995 which comprises senior officials within the department including the

Vice Chairman Joint Chief of Staff (VCJCS)[16]. Supporting the IW Executive Board is an IW Council chaired by the Assistant Secretary of Defense for Command, Control, Communications and Intelligence (ASD(C3I)). The Executive Board is chartered to address IW roles and responsibilities and serve as the DOD focal point for IW discussion at the national level. The Board is a welcome addition as it demonstrates an awareness by senior DOD officials of the need to coordinate IW, not only within the department, but in the interagency arena. However, senior officials are too busy to spend a great deal of time on any one issue, particularly one that has not reached crisis proportions. This is the point Robert McNamara makes in his book *In Retrospect* when discussing the evolution of U.S. policy in Vietnam during the Kennedy and Johnson administrations and the same is true today. [17] You need to have the pull from the top, but it is important that lower levels are fully energized.

ASD C31: The ASD(C3I), as the senior IW advisor to the Secretary of Defense, has organized a small IW Directorate comprising less than ten personnel to help him execute his responsibilities. It was the forerunner to this office which drafted the original IW directive in 1992. The IW Directorate conducts centralized planning, coordination, and oversight for IW and conducts program reviews of selected Service and defense agency IW efforts. In May 1995 the IW Directorate sponsored an IW wargame for senior government officials designed to raise the profile of the threat to the U.S. information infrastructure.[18]

The IW Directorate has also focused on initiating a DOD "Red Team" effort. This was one of the IW recommendations from the 1994 Defense Science Board report to "jump start Defensive IW."[19] Under this concept personnel knowledgeable in adversaries' offensive IW form a team to "attack" the DOD information infrastructure. Given the magnitude of the vulnerabilities, the objective would be to have this capability distributed throughout DOD and carried out at various levels and locations. This concept dovetails with the Computer

Emergency Response Team (CERT) which reacts to real world intrusions into computer systems. The Defense Information Systems Agency (DISA) has stood up small offices which accomplish both these tasks, but a more comprehensive program is necessary. In the twelve months prior to July 1994 DISA detected roughly 3,600 attacks on military networks, but officials estimate they detected only two percent of all the attacks, raising the estimated number to 182,000.[20]

The "Red Team" program is designed to increase awareness throughout DOD of the vulnerabilities of automated systems and improve the overall security posture. A comprehensive "Red Team" effort can significantly reduce vulnerabilities in the near term as many existing problems are attributed to inadequate training of operators and system administrators. As the head of a CERT team stated, "the problem...is a lack of understanding and awareness and a lack of training and technical competence on the part of the user community."[21] The "Red Team" is a laudable objective, but ASD(C3I) presently only coordinates and does not direct action. The Joint Staff has designated the Joint Command and Control Warfare Center (JC2WC) at San Antonio, Texas as executive agent to support the OSD IW Red Team effort. However, implementation of the "Red Team" concept is evolving slowly given budget constraints and manpower reductions in the Services and defense agencies.

USD(P): In 1995 the Under Secretary of Defense (Policy) (USD(P)) created an Infrastructure Policy Directorate. This office focuses on emergency preparedness and shaping the role of DOD in the protection of infrastructures, including coordination between DOD and nonDOD government, and civilian/corporate owned infrastructures.[22] These are important responsibilities which clearly fall under the heading of defensive IW. As this office matures, it will be incumbent upon the IW Council and Executive Board to insure IW activities within ASD(C3I) and USD(P) are delineated and deconflicted, if necessary.

Joint Staff: In the Joint arena, IW organizations have also emerged. On the Joint Staff, proponency for IW is now shared between the Directorate for Operations, J3, and the J6, Directorate for Command, Control, Communications, Computers, and Intelligence (C4I) under a Memorandum of Understanding signed in October 1994. While conceptually this may have merits with each directorate bringing a unique dimension of IW, in reality as Lt Gen Clapper, former Director, DIA, has recommended, the Operations Officer should be the overall staff coordinator as he is for all other operations issues.[23]

The current arrangement on the Joint Staff presents some unique challenges as no one is actually in charge. The J3 and J6 principals are too busy to dedicate the constant attention this area requires and day-to-day responsibility for IW on the Joint Staff is delegated. A need exists for direct flag officer sponsorship to orchestrate joint IW policy and doctrine development, conduct operational planning, and establish requirements. A dedicated flag officer sponsor would greatly facilitate coordination with Services, OSD, the Intelligence Community and as IW matures, the interagency and civilian sectors. It would also send a strong message that IW is an important joint warfighting issue requiring immediate highlevel attention.

Within the Joint Staff J3, responsibility for offensive IW now resides within the Information Warfare /Special Technical Operations Division (IW/STOD). This division is responsible for coordinating compartmented planning between the Services, Combatant Commands, and DOD agencies. Bob Woodward writes in *The Commanders,* the Special Technical Operations Center (STOC) is "a command and communications center for operations involving the sensitive "black" programs known only to those cleared to the special-access compartments."[24] The IW/STOD also has responsibility for coordinating all facets of C2W for the Joint Staff including policy, doctrine, and operational issues. It is this office which authored MOP 30, led the preparation of Joint Pub 3-13 and will draft IW doctrine.

This arrangement underscores the important linkage between C2W and IW. As the military proceeds to operationalize IW, the IW/STOD represents the linchpin for ensuring the integration of all dimensions of joint IW.

It has been suggested that, just as we continue to use a Single Integrated Operational Plan (SIOP) for strategic nuclear warfare, DOD might consider the use of an "IW SIOP" which addresses offensive and defensive deconfliction and intelligence equity issues.[25] If this were implemented, the task would be assigned to the IW/STOD to coordinate the task. Today the IW/STOD focuses principally on support to the Combatant Commands, but as IW matures with both its non-lethal and deterrence potentials, greater interagency participation and coordination will undoubtedly occur.

JC2WC: The activation of the Joint Command and Control Warfare Center (JC2WC) at San Antonio, Texas in October 1994 provided a valuable resource for the Commanders of the Combatant Commands (CINCs).[26] As a field-operating agency of the Joint Chiefs of Staff and headed by a flag officer,[27] the JC2WC is fully engaged in the warfighting application of IW. With 163 assigned personnel, the JC2WC dispatches tailored teams to augment CINC and Joint Task Force staffs and provide C2W expertise in all joint exercises and contingency operations. Personnel from the JC2WC have participated in U.S. efforts in Bosnia and contingency operations in both Kuwait and Haiti. Given the high turnover of personnel on CINC staffs, the JC2WC is very much in demand for its C2W expertise.[28]

The JC2WC is in the unique position of being able to cross fertilize and share C2W lessons learned between the Combatant Commands. Accordingly, the organization has played a major role in developing joint C2W doctrine and will contribute significantly to the preparation of a follow-on C2W Joint Tactics, Techniques, and Procedures (JTTP) publication. At the present time, the JC2WC is fully engaged accomplishing its assigned tasks with respect to C2W. It is only now beginning to analyze its newly assigned responsibilities as executive agent in support of the OSD IW Red Team effort. As IW evolves and DOD' s role in the larger IW arena is clarified, a natural progression will be for this organization to serve as the nucleus for a Joint IW center.

Joint COMSEC Monitoring Activity. The JCMA is another field operating agency of the Joint Chiefs of Staff having IW applications. It was created in 1993 by a Memorandum of Agreement between the Service Operations Deputies and Directors of the Joint Staff and NSA. The JCMA is charged with conducting "COMSEC monitoring (collection, analysis, and reporting) of DOD telecommunications and automated information systems (AIS) and monitoring of related noncommunications signals."[29] Its purpose is to identify vulnerabilities exploitable by potential adversaries and recommend countermeasures and corrective actions. The JCMA focuses on unencrypted DOD systems and "does not perform traditional telephone monitoring," as this function remains a Service responsibility.[30] The Joint Staff Director for Operations has been assigned primary responsibility for JCMA affairs. This facilitates coordination between the JC2WC and the JCMA. The JCMA supports both real-world operations, as well as joint exercises and DOD systems monitoring. The JCMA, more so than the JC2WC, already has the expertise to perform the Red Team mission. Rather than further diluting the already stretched resources and expertise of the JC2WC, it would make better sense to designate the JCMA as executive agent to support the OSD Red Team IW initiative.

Joint Spectrum Center. The DOD Joint Spectrum Center (JSC) was activated in September 1994 under the direction of the Joint Staff J6. The JSC assumed all the mission and responsibilities previously performed by the Electromagnetic Compatibility Center, as well as additional functions. The JSC deploys teams in support of the CINCs and serves as the DOD focal point for supporting spectrum supremacy aspects of IW. Notably the JSC assists warfighters in developing and managing the Joint Restricted Frequency List (JRFL) and assisting in the resolution of operational interference and jamming incidents. While informal coordination occurs on IW related issues between the Joint Spectrum Center, the Joint COMSEC Monitoring Activity, and the Joint C2W Center, each organization interfaces separately with the CINC staffs. No formal mechanism is yet in place to ensure the warfighters obtain a coordinated IW support package.

Combatant Commands: The focus of warfighters at the Combatant Commands remains planning and executing C2W. All of the geographic CINCs now have C2W staff officers assigned in their Operations Directorates, but organizing the diverse elements which comprise C2W is a challenge for CINC staffs. U.S. Central Command (CENTCOM) has actually physically consolidated the staff officers responsible for orchestrating electronic warfare, operational security, military deception and psychological operations into a single branch. This organization can serve as the nucleus as new IW capabilities emerge and are apportioned to the CINCs. Defensive IW poses a more significant challenge. While the CINCs can take incremental measures to unilaterally reduce vulnerabilities of their information systems, given their ultimate dependence on the national information infrastructure, defensive IW must be undertaken as part of a larger DOD sponsored initiative.

Services: Each of the Services have created or are participating in Information Warfare Centers or Activities. The Air Force leads the Services and in September 1995, the Chief of Staff and Secretary of the Air Force published the 17 page, *Cornerstones of Information Warfare* which describes how "Air Force doctrine should evolve to accommodate information warfare."[31] The Air Force was also the first to establish their Information Warfare Center (AFIWC) at San Antonio, Texas in October 1993. This was accomplished by consolidating the Air Force cryptologic support center and the electronic warfare center. The AFIWC serves as the Air Force command and control warfare executive agent with approximately 1000 military and civilian personnel assigned. The Center is subordinate to the Air Intelligence Agency closely aligning it with the Intelligence Community. The Center applies the teaming concept integrating the intelligence component with operators, engineers, communications and computer specialists, both offensive and defensive.[32] The AFIWC also has an ongoing "Red Team" and CERT effort designed to improve network security in the Air Force. The Center is collocated with the JC2WC and both organizations work closely together. Personnel from the AFIWC regularly team with the JC2WC on major deployments.

On 1 October 1995, the Air Force created its first Information Warfare Squadron at Shaw Air Force Base, South Carolina. The squadron's primary purpose will be to protect Air Force computers and communications, as well as assisting in "infiltrating an enemy's computer and communications systems."[33] The squadron will support the 9th Air Force commander who is assigned the Central Command area of operations. Eventually, the Air Force will set up more squadrons to assist air commanders responsible for other geographic areas. While the IW Squadron at Shaw AFB presently only has two officers assigned, there will be as many as 40 by August 1996, when the squadron is slated to be operational. Eventually, the squadron might grow to 85 people.[34] The Air Force recognizes it is pushing the envelope, but as General Joseph Ralston, chief of Air Combat Command, stated:

> You can sit around for another 10 years and debate about what some of the problems might be [with setting up an information warfare squadron]...but you will never know until you actually get into them and try to make it work operationally.[35]

The Navy established the Navy Information Warfare Activity (NIWA) in August 1994 to serve as their focal point for IW activities.[36] Directly subordinate to the Naval Security Group, the

NIWA is located at Fort Meade, Maryland and is closely linked to the National Security Agency.

Given the rapid pace of advancing technology the Navy has given the NIWA special authority to generate requirements and procure systems. Traditionally there has been a sharp separation in the Navy between organizations responsible for setting requirements and those charged with overseeing their acquisition. However, with new generations of computers and information systems unveiled on average about every 18 months, the Navy has adopted a more streamlined approach. As John Davis, technology adviser to the Navy's Space and Electronic Warfare Directorate, indicated, "If we fall more than one cycle behind. we could find industry putting information warfare systems into potentially hostile nations at the same time that U.S. force receive the same equipment."[37]

The Navy also has established the Fleet Information Warfare Center (FIWC) at Little Creek, Virginia from existing Fleet Deception/C2W Group assets.[38] The FIWC serves as the link between the NIWA and the Atlantic and Pacific Fleets. With personnel deployed on carrier battle groups throughout the world, the FIWC fulfills a similar mission for the Navy that the JC2WC does for the joint warfighter. The IW organizational structure created by the Navy enables the FIWC to focus on near term operational requirements, while the NIWA assumes a more long-term perspective keeping abreast of IW advances and developing and acquiring systems.

Rather than create a separate IW organization, the Marines intend to assign liaison officers to the respective Service IW centers to benefit from their efforts. The Navy and Marines have also teamed up to develop policy guidance for Navy and Marine IW/C2W operations. In February 1995, the Marine Corps Commandant and Chief of Naval Operations approved a plan which states, "Navy and Marine Corps must have a fully integrated IW/C2W capability...team must organize, train, and equip its forward deployed forces to conduct IW/C2W."[39]

The Army is the latest Service to establish an IW activity. Officially activated in May 1995, the Land Information Warfare Center (LIWA) is subordinate to the Army Intelligence and Security Command (INSCOM) but is under the operational control of the Headquarters Department of the Army, Deputy Chief of Staff for Operations (DCSOPS). As with the Air Force and Navy, the Army has closely aligned its IW effort with the Intelligence Community. The LIWA is a totally new organization in contrast to the Air Force and Navy efforts which reorganized existing activities. It is also the smallest of the three Service IW activities with a projected end strength of 50-75 personnel. The primary focus of the LIWA is to provide operational support at the Army Corps and higher levels. The Army is also coordinating closely with the Marines to assign personnel to the LIWA. Given its small size, the LIWA draws upon existing capabilities in the Army including psychological operations, electronic warfare, and operational security. The Army is in the process of institutionalizing a "Red Team" effort from existing INSCOM assets, and there is close coordination and collaboration with NSA on mutual IW efforts.

The Services' efforts during a period of serious budget constraints underscore a recognition of the importance of IW. Except for the general responsibilities delineated in the original DOD IW directive, Services have received no additional policy guidance on how to implement IW. In reviewing Service responses to IW, they have established organizations which best suit their near term needs. Manpower and funding for Service IW initiatives have been reallocated internally with the Air Force taking the most aggressive approach.

Over time Service organizations can be expected to evolve as IW matures. Today much of their focus is on implementing C2W, the military application of IW. Services have also implemented "Red Teams" to train personnel and improve security awareness of the vulnerabilities associated with information systems. Additionally, while the linkages of Service

IW organizations with the Intelligence Community is valuable, it is important IW does not become an intelligence activity. IW needs to be controlled and fully managed by the warfighters as an integral part of an overall strategy. In that regard, the recent activation of the IW Squadron by the Air Force, separate and distinct from the AFIWC, is a positive step.

Intelligence Community: Within the Intelligence Community there exists an acute appreciation of the enormous impact IW has on their efforts. Each organization, DIA, CIA and NSA has established an office to orchestrate IW related activities and satisfy the needs of their consumers. NSA, in addition to it its intelligence mission, has a unique responsibility for developing "standards, techniques, systems, and equipment" for classified information.[40] NSA has demonstrated success at protecting classified systems, but has achieved less success in the increasingly vulnerable area of unclassified computer networks where it plays a supporting role.

Under the 1987 Computer Security Act, the National Institute for Standards and Technology (NIST) was assigned responsibility for developing government wide standards and guidelines for "unclassified, sensitive information."[41] The law also directed NIST to draw upon technical computer security guidelines developed by NSA.[42] To clarify the relationship between NIST and NSA, a Memorandum of Agreement (MOW) was formalized in 1989 establishing mechanisms for implementing the Computer Security Act of 1987. The MOU has been controversial because of concerns in Congress and elsewhere that it cedes NSA much more authority than was intended under the act.[43] The act envisioned NIST requesting NSA expertise as needed, but instead the MOU has involved NSA in all NIST activities related to information security.

Protecting unclassified, sensitive information is essential in developing an effective IW architecture. Given the growing threat to our information system, NSA officials are lobbying for increased efforts to protect unclassified networks.[44] While NSA certainly has tremendous expertise and is the national authority on cryptographic protection, there are reservations about having it assume a greater role. Some perceive a conflict of interest between NSA's information protection role and its principal intelligence mission. Additionally, leadership in cryptography does not imply leadership in other areas of defensive IW. DOD has assigned the responsibility of protecting its own information infrastructure to the Defense Information Systems Agency.[45] As national defensive IW policy is formulated, the security of unclassified, sensitive information must be addressed.

Non DOD Organizations: Although defensive IW has not been officially embraced outside of DOD, several standing organizations focus on this issue. The National Communication System (NCS) was established in 1963 to coordinate the planning of national security and emergency preparedness communications for the federal government under all circumstances, including crisis or emergency, attack, recovery, and reconstitution. The NCS receives policy direction directly from the National Security Council (NSC) but is managed through the Department of Defense.[46] The NCS's National Coordinating Center for Telecommunications is staffed full time by both government and telecommunications industry representatives whose mission is to respond to both military and civil emergencies: e.g., Desert Storm, Hurricane Andrew and, more recently, the bombing of the Federal building in Oklahoma City.[47]

The National Security Telecommunications Advisory Council (NSTAC) was established during the Reagan Administration to advise the President on national security and emergency preparedness issues. This senior body is composed of presidents and CEOs of major telecommunications and defense information systems companies. NSTAC works closely with NCS.

The Federal Communications Commission (FCC) also plays a strong role in reliability and privacy issues regarding the public switched telephone network. The Network Reliability Council (NRC) was created in 1992 by the FCC to investigate the reliability of the public switched network following a series of service outages in 1991. The efforts culminated in a one-thousand page document, "Network Reliability." The FCC expanded the membership of the NRC in July 1994 and gave it a new charter. Today the NRC is composed of CEOs from telephone companies, equipment suppliers, state agencies, and federal, corporate, and consumer users.

Each of these organizations, as well as NIST, are involved in aspects of defensive IW. What is lacking, however, is an overarching framework linking these disparate efforts into a coordinated effort. Within the government someone needs to be in charge. The NCS may serve as a blueprint for this effort. Not only does it have in place representatives and links to all major government agencies, including the Intelligence Community, but it is a model for governmentindustry cooperation. To be successful, defensive IW must have the support of private industry. Not only must awareness of vulnerabilities be increased, but coordinated steps taken to reduce the risk. That is the exactly the success which NCS has achieved in the telecommunications sector. As national policy is developed for IW, strong consideration should be given to creating an organization like the NCS to serve as the focal point for this effort and having it possibly work directly for the Vice President. DOD should play a role in the organization but as an active participant, not the leader.

An Azimuth for the Future

There has been a proliferation of IW activities within the Services, OSD, joint activities and defense agencies, but up to this point it has been very decentralized. Revised policy and formal doctrine will go a long way in improving everyone' s understanding of IW and their responsibilities. Within the Office of the Secretary of Defense (OSD), one central office needs to remain as the focal point for IW. It is important that the offensive and defensive dimensions of IW are fully coordinated and it made clear to everyone in the department and in the interagency arena who has the lead in OSD. This can be helped through strong leadership by the IW Executive Board.

Similarly on the Joint Staff, a single office needs to be designated as the locus for IW related issues. The central clearing house should be the, J3, Director for Operations. While both the J6 and J2 play key roles in the areas of defensive IW and intelligence support respectively, to be successful, IW must be integrated into operations and that is the purview of the J3. Given the complexity and sensitivity of the issues involved with IW, a flag officer within the J3 should be assigned full time to oversee the development of policy, doctrine, operational planning and the IW component of the Joint Warfighting Capability Assessment. IW has enormous potential for the joint warfighter. A flag officer focusing on this issue will ensure the joint perspective and associated equities are skillfully articulated as this concept is hotly debated within DOD and the interagency process.

At the joint level, strong consideration should be given to more effectively leveraging the capabilities of the Joint Command and Control Warfare Center, Joint COMSEC Monitoring Activity and the Joint Spectrum Center. Each of these organizations is a field operating agency of the Joint Staff and brings a unique dimension to IW. However, aside from informal coordination, no formal mechanism or oversight exists to ensure they provide optimum IW support to the CINCs. For example, while the JC2WC was recently designated responsibility for the joint "Red Team" mission, the Joint COMSEC Monitoring Activity already contains the nucleus for an effective joint "Red Team" capability, with responsibility for monitoring DOD automated information systems and related noncommunications signals. At the same time, close coordination is essential between the Joint Spectrum Center and the JC2WC regarding the Joint Restricted Frequency List and its impact on the planning and conduct of electronic warfare.

The Joint Staff needs to capitalize on the full potential of each of these organizations and ensure there is unity of effort in planning and executing IW. Serious consideration should be given to creating an umbrella Joint Information Warfare Activity with the JC2WC, JCMA, JSC and any other related activities, as subordinate elements. The flag officer position now assigned to the Joint C2W Center should be reallocated to provide senior leadership at the Joint IW Center. Working directly for the Director for Operations on the Joint Staff, he can ensure the Combatant Commands leverage the full potential of IW.

While each of the Services has undertaken a different organizational response to IW, these activities are critical in laying the foundation. They serve as the nucleus upon which the Services can build their respective IW architectures. These organizations will almost certainly evolve over time, but the initial framework is in place. Training the force, particularly operators and system administrators on the defensive aspects of IW, must be tackled in the near term and enable the military to function more effectively in this new environment.

While it may be premature to specify a structure at the interagency level, the National Communication System model cannot be overlooked. Staffed with full time government and telecommunications industry representatives, the NCS serves as a microcosm of the effort that is needed to truly implement a comprehensive national defensive IW campaign. Just as within DOD, there needs to be a focal point to function as the national information assurance coordinator within the government as a whole. Chronic vulnerabilities in the NII must be addressed and coordinated actions undertaken, not only in the telecommunications and automation industries, but across the spectrum of activities involving finance, transportation, power generation and most certainly the military.

Summary

IW clearly offers enormous potential and has become a critical issue for DOD and the U.S. Government. The difficulty that currently exists is the absence of a coherent IW architecture. This paper has focused on the status of the organizational component of that architecture. While significant progress in this area has been achieved within DOD since the concept of IW was formally launched in December 1992, much more work remains to be done. Within DOD the primary focus must be on supporting the joint warfighter and in this regard the existing organizational structure must be streamlined. While offensive IW remains within DOD's realm, defensive IW is truly a national issue and must involve the private sector. The interagency arena is absolutely critical to the success of defensive IW. While DOD can be a major player in this area, it can not lead. Leadership in this area must come from the White House. IW is emerging as an inexpensive, yet effective means to directly target the U.S. homeland. The U.S. must plan for this contingency in a coherent and coordinated manner and a sound organizational underpinning is a fundamental pillar to both a DOD and national IW architecture.

Notes

1. Neil Munro, "U.S. Boosts Information Warfare Initiatives," *Defense News*, 25-31

2. U.S. Department of Defense, Defense Science Board, *Report of the Defense Science Board Summer Study Task Force on Information Architecture For the Battlefield*, (Washington: Office of the Under Secretary of Defense For Acquisition & Technology, October 1994), B-16.

3. "The Defense Science Board (DSB) is a Federal Advisory Committee established to provide independent advise to the Secretary of Defense. Statements, opinions, conclusions and recommendations in the report do not necessarily represent the official position of the Department of Defense." Defense Science Board, inside front cover.

4. Ibid., ES-2.

5. C4 Architecture & Integration Division, the Joint Staff, *C4I for the Warrior*, (Washington: Joint Staff, 12 June 1992), 1.

6. The NII is "the satellite, terrestrial, and wireless technologies that deliver content to homes, businesses, and other public and private institutions...It is the computers, televisions, telephones, radios, and other products that people employ to access the infrastructure." United States Council on National Information Infrastructure, *Common Ground: Fundamental Principles for the National Information Infrastructure*, (Washington: U.S. Department of Commerce, March 1995), 1.

7. Science Applications International Corporation (SAIC), *Information Warfare: Legal. Regulatory, Policy, and Organizational Considerations for Assurance*, (Vienna, VA: SAIC, 4 July 1995), 1-1.

8. Defense Science Board, B-3-4.

9. Joint Pub 3-13, I-4.

10. Defense Science Board, B-3-4.

11. C2W is defined as, "The integrated use of operations security (OPSEC), military deception, psychological operations (PSYOP), electronic warfare (EW), and physical destruction mutually supported by intelligence to deny information to, influence, degrade, or destroy adversary command and control capabilities, while protecting friendly command and control capabilities against such actions." Col Jim Gray, USAF, "Turning Lessons Learned into Policy," *Journal of Electronic Defense*, no. 10, October 1993, 88.

12. LT Gen James R. Clapper and LTC Eben H. Trevino, Jr., "Critical Security Dominates Information Warfare Moves," *Signal*, March 1995, 71.

13. The most recent draft of C2W doctrine states, "C2W is a warfighting application of IW in military operations and is a subset of IW." Joint Chiefs of Staff, *Joint Doctrine For Command and Control Warfare (C2W)*, Joint Publication 3-13, (Washington: Joint Chiefs of Staff, coordination draft, May 1995), I-2.

14. Ibid., I-6.

15. Ibid., I-4.

16. SAIC, A-11.

17. Robert McNamara, *In Retrospect: The Tragedy and Lessons of Vietnam*, (New York: Random House, 1995), 108.

18. The Rand Corporation wargame entitled, "The Day After...In Cyberspace" depicted Iranian computer terrorists using Internet to wreck havoc on the U.S. information infrastructure (phone system, power grid,

air traffic control) as Iran posed to invade Saudi Arabia. Neil Munro, "Infowar Disputes Stall Defense Policy"; A detailed discussion of the scenario with day by day events is contained in the article by Douglas Waller, 4446.

19. Defense Science Board, B- 11.

20. Neil Munro, "Hacker Attacks Illustrate Vulnerability of DoD War Plans," *Washington Technology,* 25 August 1994, 18.

21. Armaud de Borchgrave, "Air base no match for boy with modem," *Washington Times,* 3 November 1994, 1.

22. Ibid., A-12.

23. LT Gen James R. Clapper, 72.

24. Bob Woodward, *The Commanders,* (New York: Simon & Schuster, 1991), 327.

25. Defense Science Board, B-3.

26. Previously designated the Joint Electronic Warfare Center, under the new charter the organization's responsibilities are expanded to incorporate all the dimensions of C2W. Joint Chiefs of Staff, *Charter for the Joint Command and Control Warfare Center,* CJCSI 5118.01 (Washington: Joint Chiefs of Staff, 15 September 1994), 1.

27. Although the Director, JC2WC is open to a flag officer from each of the Services, habitually the position is filled by an Air Force general officer who is dual hatted as Commander of the Air Intelligence Agency collocated at Kelly Air Force Base, San Antonio, TX.

28. Combatant Commands also receive support from the Joint Communications Security Monitoring Activity and the Joint Spectrum Center. These specialized organizations provide personnel to the CINCs upon request.

29. Vice Admiral J.M. McConnell, Director, NSA, memo "Joint COMSEC Monitoring Activity (JCMA) Concept of Operations (CONOP)," National Security Agency, 19 July 1993,3.

30. Ibid.

31. General Ronald R. Fogelman, 1.

32. "Information Dominance Edges Toward New Conflict Frontier," *Signal,* August 1994, 37.

33. Steven Watkins, "New Era Had Humble Start," *Air Force Times,* 20 November 1995, 24.

34. Ibid.

35. Ibid.

36. Robert Holzer, "U.S. Navy Begins Information War Effort, *Defense News*, 29 August-4 September 1994, 4.

37. Robert Holzer, "Navy Eyes Single Command to Guide Info Warfare," *Navy Times*, 6 February 1995, 35.

38. SAIC, A-25.

39. "Boorda and Mundy Sign Information Warfare Guidance," *Inside the Navy*, 3 April 1995, 11.

40. Congress, Office of Technology Assessment, *Information Security and Privacy in Network Environments,*103d Cong., 2d sees., 1994, 143.

41.Ibid., 61.

42. Ibid.,146.

43. The MOU authorizes NIST and NSA to establish a Technical Working Group (TWO) to "review and analyze issues of mutual interest pertinent to protection of systems that process sensitive or other unclassified information." The TWG has six members; three from NIST and three selected by NSA. Ibid.,148.

44. Neil Munro, "Hacker Attacks Illustrate Vulnerability of DoD War Plans," 18.

45. Defense Management Review Decision (DMRD) 918, September 1992, designated the Director, DISA, as the central manager of the DII. SAIC, A-37.

46. Congress, Office of Technology Assessment, *Information Security and Privacy in Network Environments*, 61.

47. Michael Higgins (higgins@cc.ims.disamil), "NII Security: The Federal Force June 5 1995," electronic mail message to Billy Hogan (hoganbilly@aol.com), 6 June 1995.

Bibliography

"Boorda and Mundy Sign Information Warfare Guidance." *Inside the Navy*, 3 April 1995, 11.

Borchgrave, Arnaud de. "Air base no match for boy with modem." *Washington Times*, 3 November 1994, 1.

Clapper, James R., LTGen and LTC Eben H. Trevino, Jr. "Critical Security Dominates Information Warfare Moves." *Signal*, March 1995, 71.

Fogelman, Ronald R., General and Sheila E. Widnall. *Cornerstones of Information Warfare*. Washington: U.S. Department of the Air Force, September 1995.

Gray, Jim, Col. "Turning Lessons Learned into Policy." *Journal of Electronic Defense.* no. 10, October 1993, 87-92.

Higgins, Michael. (higgins@cc.ims.disamil). "NII Security: The Federal Force June 5 1995." electronic mail message to Billy Hogan (hoganbilly@aol.com, 6 June 1995.

Holzer, Robert. "Navy Eyes Single Command to Guide Info Warfare." *Navy Times,* 6 February   1995, 35

_____. "U.S. Navy Begins Information War Effort." *Defense News,* 29 August-4 September   1994, 4.

Information Dominance Edges Toward New Conflict Frontier." *Signal,* August 1994, 37.

Joint Chiefs of Staff. *Charter for the Joint Command and Control Warfare Center.* CJCSI 5118.01. Washington: Joint Chiefs of Staff, 15 September 1994.

_____. *Joint Doctrine For Command and Control Warfare (C2W).* Joint Publication 3- 13. Washington: Joint Chiefs of Staff, coordination draft, May 1995.

McNamara, Robert. *In Retrospect: The Tragedy and Lessons of Vietnam.* New York: Random

Munro, Neil. "Hacker Attacks Illustrate Vulnerability of DoD War Plans." *Washington Technology,* 25 August 1994, 18.

_____."U.S. Boosts Information Warfare Initiatives." *Defense News,* January 25-31, 1993, 1.

Science Applications International Corporation (SAIC). *Information Warfare: Legal, Regulatory, Policy and Organizational Considerations for Assurance.* Vienna: VA: SAIC, 4 July 1994.

U.S. Congress, Office of Technology Assessment. *Information Security and Privacy in Network Environments.* 103d Cong., 2d sess. Washington: Government Printing Office 1994.

_____. Information Security and Privacy in Network Environments OTA Report Summary. 103d Cong., 2d sess. Washington: Government Printing Office 1994.

U.S. Council on National Information Infrastructure. Common Ground: Fundamental *Principles for the National Information Infrastructure.* Washington: U.S. Department of Commerce, March 1995.

U.S. Department of Defense, Commission on Roles and Missions of the Armed Forces. Directions for Defense. Washington: Department of Defense, May 1995.

U.S. Department of Defense, Defense Science Board. *Report of the Defense Science Board Summer Study Task Force on Information Architecture For the Battlefield.* Washington: Office of the Under Secretary of Defense For Acquisition & Technology, October 1994.

Waller, Douglas. "Onward Cyber Soldiers." *Time,* 21 August 1995, 38-46.

Watkins, Steven. "New era has humble start." *Air Force Times.* November 1995.

Woodward, Bob. *The Commanders.* New York: Simon & Schuster, 1991.

○ A CHAPTER NOT YET WRITTEN: Information Management and the Challenge of Battle Command

Colonel Adolph Carlson
U. S. Army

INTRODUCTION

Although it has become something of an old military chestnut that "no plan survives contact with the enemy," surprisingly little has been written about how a commander modifies a plan when circumstances do not permit the formal command and staff actions associated with deliberate planning. This paper will examine two case studies to show that decision making under the pressure of ongoing operations involves a fundamentally different mental process than planning in advance of operations. Moreover, because decisions during the conduct of operations must be made in the shortest time and under the most demanding conditions, the opportunities for consultation among various command echelons are minimized, resulting in the possibility of conceptual divergence between senior and subordinate commanders. Lastly, these are problems for which information technology as yet offers no solution.

THE CASE OF FITZ JOHN PORTER

In July 1878, by order of President Rutherford B. Hayes, three distinguished U.S. Army officers were summoned to West Point, New York. The senior was Major General John M. Schofield, who had been one of Sherman's subordinate commanders during the Georgia campaign.[1] Next was Brigadier General Alfred H. Terry, veteran of campaigns in the Carolinas and Petersburg and a key figure in the 1876 campaign against the Dakota Sioux.[2] The third was Colonel George W. Getty, a thirty-eight year veteran who had participated in all of the Army of the Potomac's campaigns from Yorktown to Appomattox.[3] These officers were directed to preside over one of the most remarkable hearings in the history of American military jurisprudence, the investigation of the "facts of the case of Fitz John Porter, late Major General of Volunteers."[4]

Fifteen years earlier, a court martial had convicted Porter for his actions during the second battle of Manassas, August 1862, when he commanded the Army of the Potomac's V Corps, attached to Major General John Pope's Army of Virginia. Porter was accused of not moving his corps in accordance to orders and of failing to attack Confederate General Jackson's forces when an attack could have prevented defeat.[5] At the trial, Porter's defense argued that Pope's orders were impossible to execute because they were based upon an inaccurate picture of road conditions and the enemy's disposition. Porter could not have attacked Jackson without fighting Confederate General Longstreet's forces, which were concentrated in front of him when he received Pope's order. As evidence, Porter's side produced a dispatch from the commander of Union cavalry, Brigadier General John Buford, which reported Longstreet's troops pouring toward Porter almost eight hours prior to the dispatch of Pope's order.[6]

Shoring up the case against Porter was a body of testimony that can only be described as incompetent and immaterial. The most outlandish was the statement of Lieutenant Colonel Thomas C. H. Smith, who testified that, after meeting Porter, he had reported to Pope: "General, he [Porter] will fail you." [7] When pressed in cross examination for an explanation, Smith claimed,

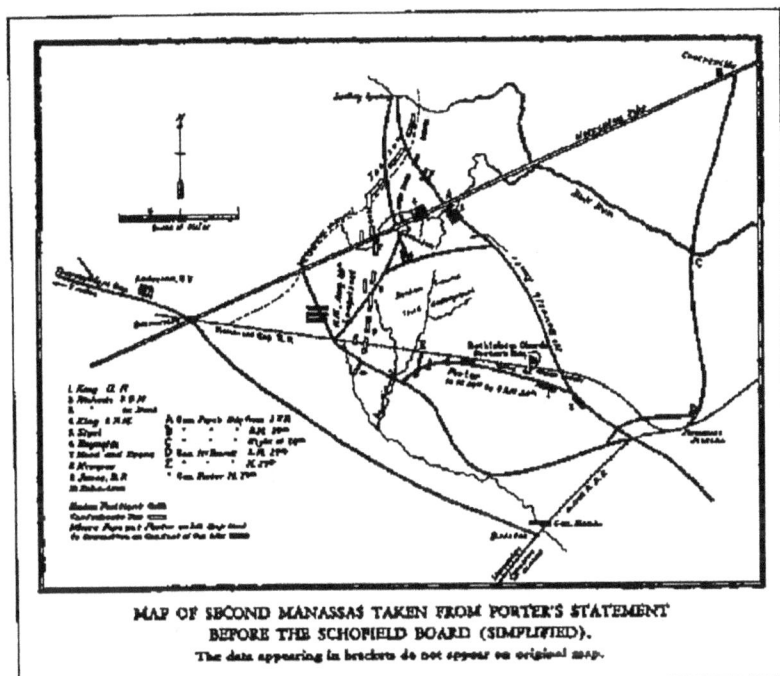

MAP OF SECOND MANASSAS TAKEN FROM PORTER'S STATEMENT
BEFORE THE SCHOFIELD BOARD (SIMPLIFIED).
The data appearing in brackets do not appear on original map.

Figure 1   Map from Eisenschiml, page 230.

I had one of those clear convictions that a man has a few times perhaps in his life as to the character and purposes of a person when he sees him for the first time.[8]

Thereafter, Porter's supporters would refer to Smith as "the mind reader."

Despite the flimsy case against him, Porter was convicted and sentenced to "be cashiered and forever disqualified from holding office of trust or profit under the government of the United States." [9] Porter appealed the verdict, but it took fifteen years for the government to act on his appeal.

The President had authorized the Schofield board to review the alleged irregularities of the 1863 court martial and to consider new evidence. Accordingly, to their great credit, ex-Confederate officers who were at the Second Battle of Manassas came forward to clarify the tactical questions on which Porter's claims rested. Most notable was General Longstreet, who revealed that at the time that Porter received orders from Pope, his Confederate troops were present in strength, and that had Porter attempted to attack "we could have broken up" the Union force and "thrown everything we had in pursuit."[10] Longstreet testified that Porter, by maintaining his position, had prevented him from joining forces with Jackson, thereby averting a greater catastrophe on the twenty ninth of August than actually occurred on the thirtieth."[11] Rather than censure, Porter's actions merited his commander's thanks. The board then called Thomas C. H. Smith, the "mind reader." Smith stuck to the version of the facts he had recounted fifteen years earlier.[12] He also told the board that he was working on a history of the Second Manassas Campaign. When Porter's counsel expressed his regrets that he would have to rewrite the portion of his history on Porter, Smith quietly said, "That chapter is not written yet."[13] The recommendations of the Schofield board were unambiguous:

In our opinion, justice requires such action as may be necessary to annul and set aside the findings and sentence of the court-martial in the case of General Fitz John Porter and to restore him to the position of which that sentence deprived him.[14]

Unfortunately, however, the board did not have the authority to grant a reversal. The case had become a political issue, and was hotly debated in Congress. Again, some of the most powerful voices in Porter's defense came from Confederate veterans. Alabama Congressman Joseph Wheeler, a former leader of Confederate cavalry, expressed the prevailing sentiment when he declared that "the honor of an American soldier was as dear to the people in the South as in any other section of the land."[15] Finally, in August 1886, twenty-three years after the original verdict, Fitz John Porter's conviction was set aside and his rank and good name were restored. The modern reader might be tempted to think that the disaster at the Second Battle of Manassas and the unjust conviction of General Porter were the results of poor communications and information management inefficiencies that have been remedied by modern technology. We might be tempted to imagine that with such innovations as space-based position locating systems, overhead imagery, near real-time battlefield information, and instantaneous communications, such a calamity could never happen again. Perhaps we had better think again.

THE CASE OF GENERAL FREDERICK FRANKS

The four days of DESERT STORM's ground operations in February 1991 seemed to most Americans to be a remarkable military achievement and a satisfactory ending to what could have been a long and bloody war. Army Chief of Staff General Carl Vuono captured the public's mood when he said:

For as long as Americans honor their history, these 100 hours of Operation Desert Storm will be remembered as one of the most powerful applications of military might and one of the most flawlessly executed campaigns in the annals of warfare.[16]

It came as something of a shock, then, when one year after the event *Army Times* writer Tom Donnelly revealed that the theater command structure was "riven by disputes"[17] over how the ground battle should be waged. Donnelly related that the theater commander and chief, General H. Norman Schwarzkopf, was often at odds with his subordinate land force commanders, Lieutenant Generals John Yeosock, Commander THIRD Army, Gary Luck, Commander XVIII Airborne Corps, and Frederick Franks, Commander VII Corps. The principal target of Schwarzkopf's frustration, Donnelly reported, was Franks, who was "not aggressive enough in attacking Iraq's Republican Guard."[18]

Schwarzkopf later added to the controversy in his October 1992 autobiography. Schwarzkopf described Franks' plan as "plodding and overly cautious."[19] Schwarzkopf told of his frustration in finding that, on the morning after the beginning of the ground attack [G+1], VII Corps had not advanced at a rate commensurate with other units in the attack, most notably the 24th Infantry Division in the adjacent XVIII Airborne Corps.[20] In the end, Schwarzkopf toned down his criticism, saying that he had been "too hard" on VII Corps' "slow progress during the battle"[21] and conceding that Franks had been "faced with the challenge of accomplishing [the] mission while sparing the lives of as many of his troops as possible."[22] Schwarzkopf closed with the thought:

> We will probably never know whether attacking the Republican Guard one or two days sooner would have made much difference in the outcome. What I did know was that we had inflicted a crushing defeat on Saddam's forces and accomplished every one of our military objectives. That was good enough for me.[23]

The fact that Schwarzkopf stopped short of indicting Franks did not deter others from building a case against him. In June 1993, retired Air Force Colonel James G. Burton, a fourteen year veteran of the Pentagon and noted critic of Army testing and acquisition, published an article in which, like the "mind reader" in the Porter case, he claimed to have had a premonitory insight into Franks' failings. Burton charged that Franks' unhurried maneuver was the result of rigid adherence to a doctrine not suited to the demands of modern maneuver warfare.[24] He argued that Franks could not keep pace with the demands of the ground war because of his overriding concern that the formations under his command remain "synchronized." The result was a linear, ponderous maneuver which permitted the Iraqi Republican Guard, Franks' objective, to escape. Burton declared that the events of the ground war proceeded at a rate "much quicker than Franks could handle,"[25] suggesting that "dinosaur blood runs freely through his veins."[26]

In a June 1994 article, retired Marine Corps Lieutenant General Bernard E. Trainor described Franks as "well respected in the Army," but "known to be slow and deliberate in all that he did . . . not what Schwarzkopf was looking for as leader of the main attack against the Iraqis."[27] Trainor concluded that Franks "could have been more aggressive,"[28] but traced the root of the problem to "a complex combination" of factors, including "different war fighting cultures" and "leadership styles."[29]

Both of these cases illustrate that different war fighting cultures can produce incompatible leadership styles. The case of Fitz John Porter suggests that this dilemma has been with us at least since the Civil War. The case of Frederick Franks warns us that even in the information age, it is a problem we ignore at our peril.

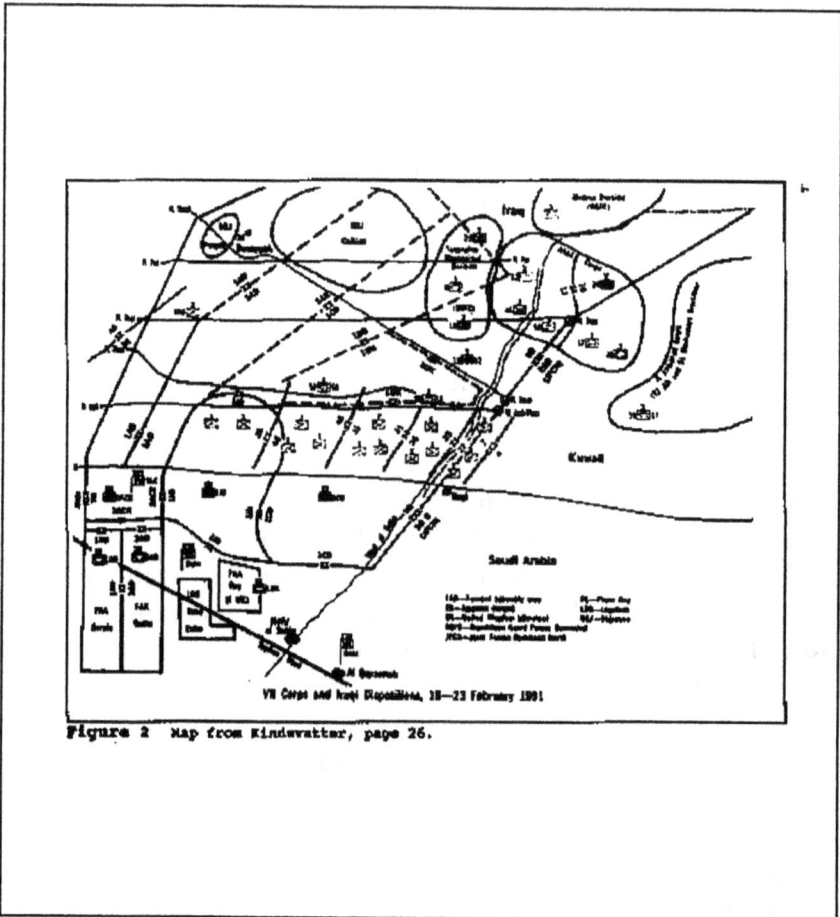

Figure 2 Map from Kindsvatter, page 26.

THE ENDURING PROBLEM OF PERCEPTION

A comparative survey of the details of the Porter and Franks cases illustrates that an organization's warfighting culture will shape a subordinate commander's evaluation of information and interpretation of direction. Moreover, when a unit is detached from one organization and placed under the operational control of another, it will carry with it the warfighting culture of its parent command. This phenomenon affects mission analysis, appraisal of enemy capabilities, appreciation for ambient conditions, and promulgation of subsequent guidance.

Mission Analysis

The missions of both the Army of Virginia and Central Command included geographic and enemy-oriented objectives. The mission of the Army of Virginia was to protect "Western Virginia and the National Capital" and to "attack and overcome the rebel forces under Jackson and Ewell."[30] Similarly, Central Command's mission required it to "eject Iraqi Armed Forces from Kuwait" and "destroy the Republican Guard."[31] Strictly speaking, in neither mission was one component more important than the other, but both Pope and Schwarzkopf chose to concentrate on the enemy-oriented aspects of their missions. Pope's orders of 27 August included the optimistic prediction that "We shall bag the whole crowd [i.e., Jackson's force]."[32] Likewise, Schwarzkopf told his subordinates "we need to destroy not attack, not damage, not surround - I want you to *destroy* [emphasis in the original] the Republican Guard."[33]

Appraisals of Enemy Capabilities

In both the Porter and Franks cases, the corps level appraisal of enemy capabilities was inconsistent with the theater commander's.

On the eve of the second battle of Manassas, Pope was under the impression that Jackson was fleeing for his life.[34] Pope's information was based on an intercepted message[35] and his own underestimation of enemy capabilities.[36] Earlier, Pope's 14 July 1862 order, calling for his command to "discard such ideas" as "taking strong positions and holding them, of lines of retreat, and of bases of supplies"[37] was a bombastic appeal to discard a cautious style of operations in favor of bolder action.[38] Porter, on the other hand, formed his judgments based on Buford's report, which indicated that on the morning of 29 August over 14,000 Confederates had passed through the Thoroughfare Gap and were massed in the vicinity of Union forces.[39] Pope denied seeing this crucial piece of battlefield intelligence until 1900 hours that evening. [40] Thus, the two commanders made decisions based on two distinct images of the enemy situation.

Similarly, Schwarzkopf viewed the enemy's collapse on G-Day as prelude to a general rout.[41] Schwarzkopf was contemptuous of the enemy facing VII Corps. He said: "The enemy is not worth shit [sic]. Go after them with audacity, shock action, and surprise."[42] To Franks, however, the indicators that suggested to Schwarzkopf that Iraqi forces were in flight painted a different picture, that they were concentrating, possibly for offensive action. Franks expected to fight five heavy Republican Guard Forces Command [RGFC] divisions,[43] four of which were estimated to be 75-100% effective on the eve of the ground attack, the fifth 50-75% effective.[44] Intelligence had warned that "[t]he RGFC is the best equipped and best trained force in the Iraqi ground forces[45] ... untainted by years of defensive warfare ... a highly motivated and trained offensive force ..."[46] During the Iran-Iraq war, the RGFC "assumed a tactically offensive role: the counterattack."[47] Like Pope, Schwarzkopf based his judgments on data provided by remote sources: intercepted messages and technical surveillance. In contrast, Franks made decisions based on battlefield data, which he believed to portray the situation with greater fidelity. Non-contextual electronic data tracking vehicular movement presented no coherent, persuasive grounds to expect that the anticipated meeting engagement would be anything less than originally anticipated.

Supervision of Operations

The manner of issuing direction in the two cases also bears comparison. Pope's orders were vague and difficult to interpret. Porter can not be blamed for failing to deduce that he was to attack from this order:

move forward ... towards Gainesville. ... as soon as communication is established ... the whole command shall halt. It may be necessary to fall back behind Bull Run ... tonight. [48]

Schwarzkopf's direction was more direct, but may have been too metaphoric. In his 14 November 1990 commanders' briefing, he directed, "I want the VII Corps to *slam* [emphasis in the original] into the Republican Guard. "[49] Schwarzkopf's memoir implies that he intended the attack to be swift and agile, but the language invokes an image of irresistible mass. Once operations were in motion, neither Pope nor Schwarzkopf conveyed their direction in person, but rather through messenger or tin Schwarzkopf's case] by electronic means. For Franks, the additional guidance did not clarify. On 25 February, Schwarzkopf called the VII Corps command post and talked to Colonel Stan Cherrie, corps G-3 [Franks was forward with the 3d Armored Division]. Schwarzkopf reportedly told Cherrie to keep pressuring the enemy: "I want you to keep the Bobby Knight press* on them."[50] This was another example of metaphoric language intended to be emphatic, but which was too imprecise to convey intent. Not until 26 February was the more specific direction passed to VII Corps, to change the operation from "deliberate operations to a pursuit."[51]

Appreciation of Ambient Conditions

Common to the specifications charged against Porter and the deficiencies alleged against Franks were the difficulties of night operations. In Porter's case, Pope had directed Porter to conduct a night march on 27 August, commencing at 0100 to arrive at Bristoe Station by daybreak.[52] Porter could not obey the letter of this order because of factors Pope could not appreciate: the road

---

* A "press" in basketball is a term used to describe a defense which takes the initiative away from an offensive team by coordinated action all across the court. Bobby Knight, Indiana University basketball coach, was famous for employing this tactic.

was narrow, Confederate forces had destroyed the bridges, and two to three thousand Union Army wagons blocked the way.[53] Porter did not start his march until 0300 and could not reach his destination until 1000, a delay which Pope maintained had prevented him from "bagging" Jackson's forces.[54]

In Franks' case, Schwarzkopf's anger at what he perceived to be VII Corps' failure to "make good progress during the night"[55] of G-Day suggests that, like Pope, he could not visualize his subordinate's predicament. Because of his wide span of control over U.S. and coalition forces, Schwarzkopf's staff relied on computer graphic displays of information, which were necessarily abridged.[56] These displays precluded him from absorbing the details of any specific component of the operation, even in the area of the main attack. Since his map showed only movement, he was inclined to think that a lack of movement equated to a lack of progress. In his own words, "They seem to be sitting around."[57] He could not appreciate the difficulties of the 1st Infantry Division's consolidating a breachhead, the passage of 7000 vehicles of the British 1st Armored Division, and then the 1st Division's redeployment to join 1st and 3d Armored Divisions.[58] For all the advantages of twentieth century technology, Schwarzkopf had no better picture of Franks' situation than Pope had of Porter's.

THE PHENOMENON OF CONCEPTUAL DIVERGENCE

Although separated by over a century, both of these cases illustrate a common problem: diverging concepts of ongoing operations leading to dysfunctional misunderstandings at different levels in the chain of command. This divergence is a natural consequence of on-the-spot decision making when conditions preclude consultation and coordination. A model might be useful in understanding the phenomenon.

Prior to the initiation of active operations, deliberate planning should produce a common vision between senior and subordinate commanders. The record shows that Pope and Porter agreed on the plan to concentrate the Army in the vicinity of Alexandria to confront Lee's threatened move north.[59] Similarly, Schwarzkopf and Franks appear to have been in general agreement about the concept of the "great wheel" to defeat the Iraqis.[60]

After the senior and subordinate reach consensus, each begins a process of subsequent decision making independent of the other. Once combat operations commence, decisions must be made at a rate that does not permit the formal, fully-staffed process, especially when the demands of supervision compete with the demands of decision making. Accordingly, commanders involve a smaller number of staff officers and make decisions using on abbreviated procedures. These procedures will vary according to the commanders' personalities, but will include the following analytical processes:

- detailed planning, to ensure that each of the mission's component tasks is assigned to the most capable element.
- fine tuning, to provide refined guidance to subordinate units based upon the most detailed data available.
- contingency analysis, so that the organization is prepared to respond effectively to possible changes in the enemy or friendly situation or to unpredictable variations in terrain or weather.
- disaster avoidance, to avoid catastrophic defeat in the event of the worst case.
- updating, to ensure that actions underway or contemplated are still appropriate to the current situation.

In the absence of any other variable, there is already a likelihood that a subordinate commander's concept of operations will diverge from his superior's, because regardless how much we enhance the collection and dissemination of data, the ability to arrive at a decision based on that data can not be automated. If the two are making their assessments based on data from different sources, that divergence is likely to be more pronounced, because the commanders will view each piece of data in different contextual settings.

THE CRITICAL RELEVANCE OF CONTEXT

The transmission of data without the associated context further diminishes the clarity of a message, especially when the receiver is in a different contextual environment than the sender. The more impersonal the means of transmission, the greater the lack of context will produce misinterpretation. Any alternative to face-to-face consultation reduces the ability of senior and subordinate commanders to communicate clearly. The use of non-specific or metaphoric language carries with it the greater risk that the image the sender intends to communicate will not match with the image invoked in the mind of the receiver.

As more powerful technological tools intrude into the process of command, they bring with them the risk that a generation of officers will be more inclined by instinct to turn to a computer screen than to survey the battlefield, and that the use of precise operational terms will be displaced by computertalk. If that happens, we may have lost more than we have gained. To use Clausewitz' terms, a commander's ability to perceive "some glimmering of the inner light which leads to the truth,"[61] may be enhanced by technology, but the "courage to follow this faint light wherever it may lead"[62] is still a function of character.

THE ROLE OF INTERMEDIATE COMMANDERS

Significantly, in both cases there was a level of intermediate command between the corps and theater commanders.

In Porter's case, Major General Irwin McDowell, who was senior to Porter, assumed command of his own and Porter's corps, in accordance with the custom of the day.[63] McDowell had been with Pope since the Army of Virginia was created, and enjoyed his confidence.[64] As McDowell positioned the two corps, Porter looked to him for clarification of Pope's intent. "What do you want me to do?" Porter asked McDowell at a critical point.

McDowell only waved his hand and rode off, leaving Porter to rely on his own judgment.[65] McDowell's absence was a subject of testimony at Porter's court martial. Porter stated that: "From about 10 A.M. ... till after 6 P.M., I received no instructions from him [i.e., Pope] or General McDowell, though I had sent many messages to both of them."[66]

In Franks' case, the intermediate commander was Yeosock, Commander of the THIRD Army. In accordance with modern U.S. military doctrine, Army command in a theater of operations involves three distinct responsibilities: to provide Army forces to the joint force, to perform assigned combat support and service support functions in the theater, and to provide command and control over Army elements engaged in operations.[67] In the case of THIRD Army, plans and exercises leading up to the deployment into Saudi Arabia emphasized the first two responsibilities, but the command and control of operational Army elements was performed primarily by XVIII Airborne Corps.[68] Thus, when the President decided to reinforce CENTCOM with VII Corps, THIRD Army was forced to assume a function for which it had not prepared. Both McDowell and Yeosock occupied posts which required either that they advocate their subordinates' views to the superior or that they compel their subordinates to alter operations in accordance with the superiors' guidance. Neither intermediate commander performed effectively, because both chose to be distant from and incommunicative with their subordinates. For both Porter and Franks, this remote style of command may have been the most difficult aspect of their new environment because it represented a major cultural change. Porter had been a protege of McClellan, a general who made a practice of "long days in the saddle and nights in the office - a very fatiguing life, but one that made my power felt everywhere and by everyone."[69] Similarly, the command style in U.S. Army Europe during Franks service in it was summarized by its commander, General Crosbie E. Saint:

When you personally talk to commanders, things come out that you cannot get from a telephone conversation. I have no doubt about the need for that kind of personal coordination. That is the reason why a corps or army group commander* needs a mobile command post. The commander can send it out ahead of time to someplace convenient and then bring commanders together to get everything synchronized.[70]

In another article, Saint described the dangers of overreliance on computer screens and teleconference as "green table" mentality.[71] The reference comes from a quote from von Moltke, "War cannot be conducted from a green table." Saint asserted that "remote control of land operations remains an illusion."[72] In contrast, Schwarzkopf related that his job was to "stay in the basement with our radios and telephones, assessing the offensive as it developed ..."[73] If the span of Schwarzkopf's military and political responsibilities tied him down to the war room at Riyadh, Yeosock's immobility is more difficult to justify—communications were operating reliably, and the air situation would have facilitated helicopter movement among his subordinate commanders' headquarters. Had he been closer to the action, he could either have argued credibly that Franks' judgments were sound, or alternatively pressed Franks to execute in a manner more in line with Schwarzkopf's concept.[74]

Significantly, neither McClellan nor Saint felt the criticism leveled against their subordinates was justified. McClellan called Porter "probably the best officer general officer I had under me."[75] Likewise, Saint lauded Franks' "incredible success."[76]

---

*Author's note: An army group is a NATO command echelon, analogous to a field army commander in the CENTCOM example.

THE MORALE FACTOR

Another factor that Schwarzkopf might have considered before publishing his memoirs is the impact of his criticism on the morale of the rest of the officer corps. He might have found an insight in the papers of Major General George Gordon Meade. Meade made a number of references to the Porter court martial. In March 1863, Meade related that he had been called to Washington to appear before the "Committee on the Conduct of the War,"[77] to comment on the conduct of Major General William Buell Franklin, who, like Porter, was investigated for the defeat at Fredericksburg. Meade thought that Franklin was being done a "great injustice[78] ... to be made responsible for the failure at Fredericksburg ..."[79]

This practice of blaming scapegoats caused much bad feeling among the officers of the Army of the Potomac. Meade himself was affected. In a 29 June 1863 letter to his wife] he revealed that when the War Department messenger arrived at his tent to inform him that he was to command the Army, his first thought was that he was about to be arrested, 80 likely to take the blame for Chancellorsville.

The practice of blaming failure on subordinates was one of the least appealing characteristics of the Army during the Civil War, but while we may not condone it, we can at least appreciate that it was a result of repetitive defeats. It seems much more unseemly in the wake of victory, regardless of how far short that victory fell of the commander's claims. Although the **Gulf War** ended before the decline of morale that characterized the Army of the Potomac could become a factor, the question nonetheless remains: what will be the impact of Schwarzkopf's criticism of Franks on the officer corps as a whole?

CONCLUSION

The Porter-Franks comparison cannot as yet be neatly concluded. Porter fought for over twenty years to have his conviction reversed, but in the end was exonerated. Franks was never indicted, so his reproof takes the form of faint praise in Schwarzkopf's memoir and pettifoggery from the likes of Burton. Since the Iraqi generals are never likely to be as gracious in defeat as the Confederates were, we will never know whether more rash maneuver on the part of VII Corps would have resulted in an American unit falling into a carefully laid Iraqi ambush. As long ago as the fourth century, the military commentator Vegetius warned that:

> A rash and inconsiderate pursuit exposes an army to the greatest danger possible, that of falling into ambuscades and the hands of troops ready for their reception.[81]

And while the evidence suggests that many RGFC formations escaped destruction, their escape now appears to be a result of the decision to end the war after four days of ground combat, inefficient employment of fixed and rotary wing aircraft, and less than optimal terms of the cease fire—factors that were not the result of anything that Franks did or failed to do.

The amalgamation of information technology with the proven techniques of effective operational command is a chapter not yet written in the literature of information-age warfare. Franks himself has taken the first steps as commanding general of the Training and Doctrine Command. The concept of "battle command"[82] in the latest Army operations manual, written under Franks' direct supervision, is a product of his reflection on the subject. More than any of the other major figures in the war against Iraq, Franks has refused to rest on his laurels and has endeavored to provide the army the benefit of his experience.

Franks' actions are reminiscent of another corps commander in another war, Major General Hunter Liggett, commander of the American Expeditionary Force's I Corps in the Argonne Forest, in World War I. On 12 November, 1918, twenty four hours after the Armistice, General Pershing visited the I Corps headquarters, and found the corps commander poring over his maps. "Don't you know the war's over?" asked a bemused

Pershing. Liggett replied, "I'm trying to see where we might have done better."[83]

Notes

1. Webster's American Military Biographies [Springfield, MA: G.& C. Merriam Company, 1978] page 369.

2. Ibid., page 433.

3. George W. Cullum. Biographical Register of the Officers and Graduates of the United States Military Academy [Third Edition] [Boston: Houghton, Mifflin and Co., 1891] pages 41-43.

4. Otto Eisenschiml, *The Celebrated Case of Fitz John Porter* [Indianapolis: Bobbs Merrill, 1950], page 214.

5. The details of the charges and specifications can be found in Theodore A. Lord. A summary of the Case of General Fitz John *Porter* [San Francisco: 1883], pages 15-16.

6. Eisenschiml, page 89-90.

7. Ibid., page 94.

8. Ibid., page 95.

9. Ibid., page 160.

10. Ibid., page 222.

11. Ibid., page 222.

12. Ibid., page 325.

13. Ibid., page 236

14. Ibid., page 249

15. Ibid., page 282.

16. Quoted in AUSA Institute for Land Warfare, *The U.S. Army in Operation Desert Storm* [Arlington, AUSA, June 1991], page 24.

17. Tom Donnelly, "The Generals' War," *Army Times*, 2 March 1992, pages 8, 16-18.

18. Ibid., page 16.

19. H. Norman Schwarzkopf, *It Doesn't Take a Hero* [New York, Bantam Books, October 1992], page 433.

20. Ibid., pages 455-456.

21. Ibid., page 482.

22. Ibid.

23. Ibid.

24. James G. Burton, "Pushing Them Out the Back Door," *Proceedings*, June 1993, pages 37-42.

25. Ibid., page 39.

26. Ibid., page 42.

27. Bernard E. Trainor, "Schwarzkopf and His Generals," *Proceedings,* June 1994, page 46.

28. Ibid., page 47.

29. Ibid.

30. Order of President Lincoln dated 26 June 1862, quoted in *The War of the Rebellion,* A Compilation of the Official Records of *the Union and Confederate Armies.* [Hereafter cited as *The War of* the Rebellion].[Washington, Government Printing Office, 1885], Series I - Volume XII- Part III, page 435.

31. United States Department of Defense, *Final Report to Congress* [Title V Report, April 1992], [hereafter cited as the Title V Report], page 73.

32. Vincent J. Esposito, *The West Point Atlas of American Wars. Volume I. 1689-1900* [New York: Praeger, 1959] map 60.

33. Schwarzkopf, page 381.

34. Eisenschiml, page 51.

35. Ibid., page 46.

36. Ibid., page 49.

37. Quoted in *The War of the Rebellion,* page 474.

38. Henry Gabler, *The Fitz John Porter Case: Politics and Military Justice.* [City University of New York Ph. D. Thesis, 1979], page 52.

39. Ibid., page 57.

40. Ibid., pages 89-90.

41. Schwarzkopf, pages 452-453.

42. Schwarzkopf, page 433.

43. Title V Report, page 256.

44. Ibid. In addition, four regular Iraqi armored divisions [the 10th, 52d, and 12th, rated at 50-75% effective, and the 17th, rated at 75-100% effective] were identified either in or in reinforcing distance of the VII Corps' zone of action.

45. National Training Center Handbook 100-91, *The Iraqi Army - Organization and Tactics* [National Training Center: S2, 177th Armored Brigade, January 1991], page 26.

46. Ibid.

47. Ibid., page 25.

48. Eisenschiml, page 55.

49. Ibid., page 433.

50. Rick Atkinson, *Crusade* [New York: Houghton Mifflin: 1993], page 421.

51. Richard M. Swain, *Lucky War* [Ft. Leavenworth, KS: U.S. Army Command and General Staff College Press, 1994], page 252.

52. Eisenschiml, page 47.

53. Ibid.

54. Ibid., pages 82-83.

55. Schwarzkopf, pages 454, 455.

56. Swain, page 105.

57. Schwarzkopf, page 455.

58. Peter S. Kindsvatter, "VII Corps in the Gulf War: Ground Offensive" *[Military Review,* February 1992], pages 24, 28.

59. Eisenschiml, page 47.

60. Schwarzkopf, page 383.

61. Carl von Clausewitz, *On War* [Princeton: University Press, 1984], page 102.

62. Ibid.

63. Eisenschiml, page 56.

64. Ibid., page 49.

65. Ibid., page 58.

66. Eisenschiml, page 59.

67. FM 100-16, *Army Operational Support* [Washington, D.C.: Headquarters, Department of the Army, February 1995], page 2-14.

68. Swain, page 9.

69. George B. McClellan, *McClellan's Own Story* [New York: Webster & Co., 1887], page 69.

70. Croshie E. Saint, "A CINC's View of Operational Art," *Military Review,* September 1990, page 76.

71. Croshie E. 1991, page 21.

72. Ibid.

73. Schwarzkopf, page 452.

74. Swain, page 334.

75. Quoted in Gabler, page 13.

76. Message, General Saint to Generals Galvin and General Sullivan, 260800Z March 1991, subject: Thoughts on the Victory in Desert Storm, page 2 of 4.

77. George Meade, *The Life and Letters of George Gordon Meade* [New York: Scribners, 1913], page 357.

78. Ibid., page 359.

79. Ibid., page 360.

80. Ibid., page 11.

81. Vegetius, *DE RE MILITARY* [Contained in T. R. Phillips(ed.) *Roots of Strategy* {Harrisburg PA: Stackpole Books, 1985}], page 166.

82. FM 100-5 *Operations* [Headquarters, Department of the Army, June 1993], pages 2-14, 2-15.

83. This incident is related in Geoffrey Perret's *There's a War to be Won* [New York: Ballentine Books, 1991], page 7.

UNINTENDED CONSEQUENCES OF JOINT DIGITIZATION

Lieutenant Colonel Steven J. Fox
U.S. Army

ABSTRACT:

The use of digital technology for future military operations will bring unintended consequences that will profoundly affect the art of warfighting. Joint digitization is an architecture that improves joint C2 functions through the availability of real-time situational information, links between sensors to shooters, and the use of integrated shared knowledge by automatically generating informational data bases. The idea of joint digitization is to enhance the warfighters C2 decision-cycle through a seamless integrated digital information network. Although it is expected technology will continue to bring huge payoffs, military professionals cannot assume that the use of technology by itself will be the panacea to achieving risk-free operations. Arguably, embracing digital technology also can bring accidental consequences that can damage and weaken a military organization. This paper is not about the science of digital technologies, but about the unplanned effects it might have upon the art of war. There are three possible unintended consequences of digital technology: 1) the merging of operational and tactical levels of war; 2) the general diminishing of a commander's prerogatives; and 3) an increase in the fragility of the force.

125

Introduction

The use of digital technology for future military will bring unintended consequences that will profoundly affect the art of warfighting. Technology has always had a deep effect on organizations that conduct warfare and, in each case, innovations have forever changed the nature of warfare. Although it is expected technology will continue to bring huge payoffs, military professionals cannot assume that the use of technology by itself will be the panacea for achieving risk-free operations. Technology, specifically digitization, promises to reduce the uncertainty and increase a warfighter's lethality; however, applying science to solve problems often leads to unanticipated new problems. Arguably, embracing digital technology may also bring accidental consequences that can damage and weaken military organizations. This paper is not about the science of digital technologies, but about the unplanned effects digitization might have upon the art of war.

Digitization is the application of micro-processors to achieve a seamless information flow for coordinating and employing war fighting assets. This paper addresses Twenty-first century issues assuming that a fully joint interoperable C4I system has been achieved, such as the objective version of Global Command and Control System (GCCS).[1] This paper also assumes that the goals of the Navy's Copernicus and the Army's Digitized Battlefield future architectures are realized and fully interoperable, as well as the intent of the Air Force's Horizon strategy.[2] The essence of this metamorphosis toward "joint digitization" is an architecture that improves joint C2 functions through the availability of near-real-time situational information, the links between sensors to shooters and the use of integrated shared knowledge by automatically generating informational data bases. It is a digitization of the entire joint battle space that processes, displays, and transfers information between echelons horizontally, inter-service, and among allies. The idea is to enhance the warfighters C2 decision-cycle through a seamless integrated digital information network that also supports the warfighter's weapons systems.[3] There are three possible unintended consequences of digital technology that will affect military leaders' approach toward the art of war: 1) the merging of operational and tactical levels of war; 2) the general diminishing of a commander's prerogatives; and 3) an increase in the fragility of the force.

Today, the operational level serves as a link between strategic aims and tactical employment of forces. Digitization unintentionally will have the effect of dismantling hierarchical structures and increasing the importance of the link between strategic and tactical levels. The traditional strategic, operational and tactical levels will gradually mutate into a blurry, flatter, two-tier hierarchy where the operational and tactical levels merge. It is not clear whether the effects of merging the levels are beneficial or detrimental to a theater of operations.

The second unintended consequence is the trimming back of the traditional boldness and initiative of subordinate leaders. Since the future portends that threats will be less well defined, military missions will have more acute, direct political consequences that will demand a tighter reign on commanders. As a result of mass media coverage, civilian leaders will pressure military leaders for constant updates in order to maintain public support. It is quite possible that the initiative and boldness traditionally expected of American warfighters will be severely constrained by a superior's strong oversight. Digitization, meant to give commanders greater autonomy, might actually strip commanders—especially those considered at the tactical level—of their individual prerogatives.

Finally, as our services improve their lethality through digitization, which also results in better situational awareness and quicker reaction times, the fragility of the force might increase. The new lack of robustness due to more complex electronic systems might negatively effect the ability of the force to quickly recover from the rigors of war. Issues concerning replacement of high-tech personnel casualties, availability of complex repair parts and ease at which long haul communications can be disrupted could all contribute dramatically to the future endurance of the joint digitized force.

Merging Levels of War

The American public's lack of stamina to endure a protracted conflict and their abhorrence of battlefield casualties requires quick, decisive, strategic results. It is the warfighter's quest for immediate strategic results that will force the restructure of the three levels of war. With the help of information technology, joint commanders of the Twenty-first century will be able to efficiently command and control geometrically greater battle spaces at a far greater tempo and confidence level. The amount of uncertainty will be less. The virtually seamless flow of information from the lowest echelons--conceivably individual battalions, ships or aircraft--gives joint commanders a significantly increased knowledge base from which to coordinate, synchronize, and employ forces. This knowledge will be shared nearly simultaneous throughout the chain of command. The force multiplying effect of shared information will be to achieve a greater coherence and unity of effort with significantly fewer casualties. Consequently, actions at every level will instantaneously effect the other and will have the affect of reducing the time between decisions.[4] Furthermore, the trend toward more powerful weapon systems will give rise to tactical elements being employed more often in direct support of strategic objectives. Situational awareness will permit the conduct of simultaneous offensive operations allowing the capability to convert tactical success immediately into decisive strategic results.[5]

In the pursuit of more decisive operations, a great shift in the power to make operational and tactical decisions at one echelon will occur. The desirability of this shift will depend on how senior leaders adapt to the greater responsibility. A goal of joint

128

digitization is to provide a greater assimilation of more information into the warfighting process. This implementation goal will ultimately expand the strategic level's span of control with the streamlining of the operational chain of command. Today, military organizations are generally sized by how much one commander can effectively control. By design, each echelon's commander is better informed to make appropriate decisions concerning his responsibilities. During the late 1980's and early 1990's, an evolution in business management led by the integration of information systems caused an increase in the span of control for a typical senior manager; this subsequently resulted in a flattening of the management structure. Likewise, digitization will provide more senior leaders the information that will permit them to feel comfortable making more judgments that were traditionally left to operational and tactical commanders. As in business, digitization will cause an increase in the span of control at the strategic level and a subsequent greater centralization of command. In the future, digital technology may render our current structure obsolete.

A flatter structure may not be always beneficial. The disadvantage of centralization is the magnitude that each decision potentially has. Mistakes in simple judgment can lead to greater, longer lasting consequences. Since the influence of a strong commander permeates throughout, there may be cases where trusting a single individual's perceptions may be, in the end, very costly in terms of resources, lives, and continued public support.

Another important reason for the merging levels of war is that communication between all echelons is expected to shift dissemination and collection of intelligence, targeting, and other data from hierarchical to a non-hierarchical command structure (See Figure 1.)

Figure 1
Command Information Structures[6]

Hierarchial Structure   Non-Hierarchial Structure

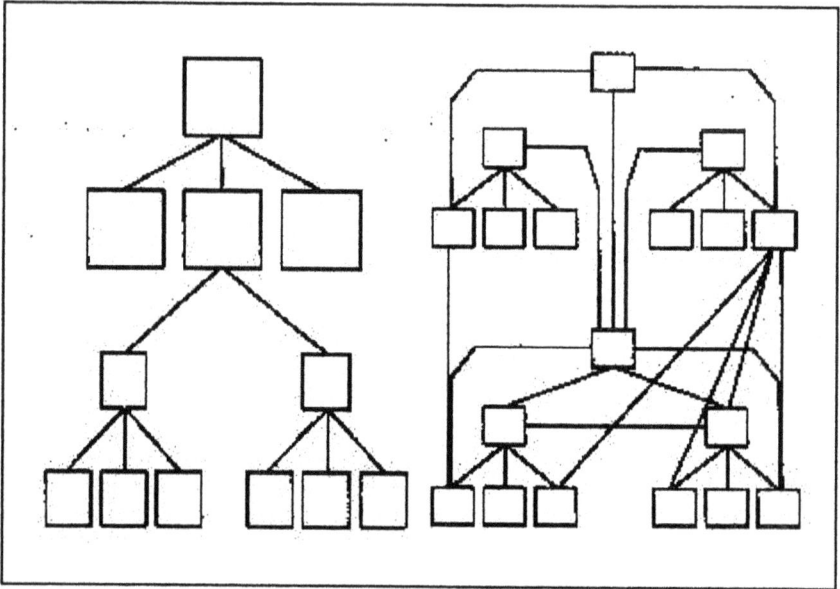

This internetted command structure undesirably leads to an "diffused command authority." [7] Even today there are indications of just how future technology can alter command approaches within the military. Electronic mail (E-mail) has created a formal, as well as, informal dissemination structure within the Department of Defense. The ease at which an individual can send a courtesy copy or a message to anyone, regardless of position, has placed unusual demands upon an organization's hierarchy. It is easy to interpolate how the future and its promise for "smarter" systems, which are better able to assimilate and display information, can change the traditional needs for hierarchical levels. What is not clear is the effect that these informal lines of communications will have on a command structure during times of conflict.

As technology increases the strategic echelon's span of control, there will be a predictable gravitation to coordinate tactical weapons at the highest possible level; thus, more pressure to flatten or merge the levels of war. Improved war-fighting capabilities which have increased the mobility, dispersion, lethality, and tempo will drive control measures at least to a joint theater level. Centralized control is needed for synchronization of combat power in order to conserve assets, and prevent fratricide and civilian collateral damage. Weapon systems could be enabled or disabled remotely whether they are on a ship, land or aircraft. It will be possible to control tactical fires and the maneuvers of combat elements to optimize the direct obtainment of the strategic goal. These combat elements will be armed with far more sophisticated direct and indirect munitions with ranges that exceed line of sight. Military professionals saw just a glimpse of technology during Desert Storm. By Twenty-first century standards, Desert Storm will look antiquated, but it illustrates just how complex and how important synchronizing forces are in order to prevent friendly casualties and to insure decisive strategic results.

The significance for future military organizations is that traditional tactical and operational levels will have the same information and same battle space concerns. The need for separation based upon capability to command and control will likely cease to exist in the Twenty-first century. The overall disadvantages of this merging has yet to form. How the senior leadership adapts and structures the resulting organizations will determine the magnitude of the drawbacks.

Diminished Command Prerogatives

Like many problems, the solutions often have undesirable consequences. It is quite possible that digitization upon Twenty-first century military organization will be detrimental to the concept of execution at the lowest level and could lead to overcontrol. With the shared vision resulting from digitization, every one within the chain of command could have access to the same situational information. Theoretically, everyone from the White House, the Joint Staff and CINC Staffs, down to the tactical levels could have access to the same facts. Potentially it could become normal for some operational and tactical level decisions to come from the White House or the Joint Staff. The United States military requires leaders that possess boldness and initiative to act on one's own judgment. These traits within our leaders make the U.S. military a formidable force. Digitization may unintentionally affect the relationship between echelons by limiting the organization's initiative, ingenuity, and inventiveness through constant monitoring by superiors.

No one likes to have someone looking over their shoulder as they work. Increased theater awareness and the ability to automatically transfer databases will result in the expansion and availability of on-line information. This may very likely result in a great deal of second guessing by senior commanders and staff. Invariably during a crisis, a subordinate commander who is closer to the action will select a different course of action than that of the senior commander. Should the superior commander second guess and redirect the effort or sit on his hands and wait it out? The challenge for senior leaders in the future will be curtailing their inclinations and allowing subordinate leaders their prerogatives. When a commander thinks he has all the facts, allowing a subordinate his prerogatives by not interfering goes against the "zero defects" culture that permeates the military today; unfortunately, there is no evidence that the future culture will be any different.

Information technology will also find it hard to digitize what is in the commander's head and what he feels--the kind of intangible awareness that comes from being close and near the area of operation. Our forces will fight the way they train. There is a danger that factors that cannot be quantified will be disregarded and our commanders not near the action will be "partly conditioned" by the technical tools available to them.[8]

For example, during a peacetime exercise, based on intuition or other undigitizable information, a commander may want to deviate from a plan during its execution. How many times will a commander accept corrections or challenges, by a "all-knowing" higher commander or superior's staff. Echelons above may believe they understand and have the "correct" bigger picture and demand adherence to previously coordinated plans or only minor tweaks to an existing plan. There are very few commanders, believing they have the right situational information, that can allow a subordinate to act contrary or independently to their expectations without at least making an inquiry. Sure, the subordinate commander could explain and defend his actions and even prevail; however, a sense of autonomy, pride and creativeness is predictably lost by a subordinate each time it occurs. It doesn't take many real-time corrections or constant challenges by a boss to stifle boldness and initiative. It might be hard for today's leaders to appreciate the loss of initiative and boldness caused by digitization, since they have had these traits nurtured into them throughout their careers. The concern should be about the generation of leaders who grow up with extensive and perhaps constant oversight and monitoring.

The Marines "Warfighting" manual, FMFM 1, has recognized this concern and has succinctly stated, "Equipment that permits over control of units in battle is in conflict with the Marine Corps's philosophy of command and is not justifiable."[9] The digitization technology of the Twenty-first century, by obtaining massive situational awareness, may have the unintended consequence of permitting commanders too much control. The military services could be entering into a new era of electronic micro-management.

The pressure to micro-manage, to ask questions, and to second guess field commander's decisions will be very high. Consider a CINC or a Joint Task Force commander, with the military and civilian leadership in Washington D.C. and the media doggedly pursuing in "real time" their individual stories, demanding to know status and planned intentions. The pressures will be enormous for information. With instantaneous civilian global information networks, such as CNN and broadcast network news, our civilian population has become more addicted to war news.[10] For the future, the "CNN war leads public and leaders to define political events in terms of the video clips and sound bites that compose TV news images."[11] A causal effect of this coverage is that the available reaction time to events for civilian policy makers is going to decrease because of amplified public interest.[12] The American people are going to demand more information from their senior civilian and military leaders. Accordingly, policy makers are going to want to know more and want to know it sooner.

The pressure upon senior military leaders to supply only the most accurate information and not to publicly embarrass themselves, as well as their civilian leaders, will be immense. The future joint leader and corresponding staff are going to be forced to make larger number of inquiries to satisfy the media and public. Situational awareness through digital technology will be the enabling technology for maintaining public support for the military, as well as keeping the political leadership from embarrassment.

A good counter argument might be that unintrusive electronic queries will be extensively used and the distraction will be very minor. This argument holds well, if the question is strictly qualitative in nature, is an acceptable answer, or doesn't require explanations. Unfortunately, it is more likely that one answer will precipitate several more questions that may or may not be available unobtrusively through an electronic data base. Digitization might inadvertently and unanticipatedly increase outside distractions through the answering of endless questions up the chain of command. On the positive side, it is easy to see how this kind of uncontrolled eavesdropping by a higher level staff can lead to better situational awareness and better serve a joint commander. Sadly, it is hard to imagine how unencumbered, direct access to planning and execution information by a superior and his staff, without the subordinate's personal spin, could have a positive effect on the hard-charging, self-motivated commander.

One could argue that the higher echelons will not be interested in information that is specific and in detail. It is more likely that the smaller Twenty-first century force structure will limit the options available to our senior leaders. Consequently, leaders will require more specific information in order to make decisions. For example, the Joint Staff is currently in the process of developing and fielding the first phase of the Global Command and Control System (GCCS). ADM Owens, Vice Chairman, Joint Chief of Staff was given a demonstration of the Global Status of Resources and Training System (GSORTS). While being impressed with its capability to easily understand air base and other readiness ratings, he indicated it would be nice to also receive the aircraft availability by individual weapon's load, including Precision Guided Munitions (PGMs), and by sensors in a particular area of responsibility.[13] What is significant is not that the Admiral wanted to know this information, but how the information going to be used at the strategic level. What if the information is not as expected or quantifiably correct? Very few leaders like to be surprised, and with the relatively smaller force structure, specifics are becoming very important.

Another concern is whether our warfighters will be prepared for the avalanche of questions. It cannot be discounted that routine questions from higher echelons tend to increase exponentially in importance the farther down the chain of command they propagate. So the lower echeloned, smaller staffed, war-fighting organizations are less prepared to handle queries and are distracted disproportionately.

Often in discussions about digitization and automation, it is surmised that staff sizes will decrease in numbers.[14] It is quite possible that staff sizes may need to increase--not decrease. The requirement for information is compounded by the problem that during peace there is a tendency to become over organized and more bureaucratic in structure.[15] This tendency can be summarized as the "need-to-know everything" syndrome. The propensity to explain and track data is gradually taking up more of the commander's and staff's time. As our services become more complex, more records and information are required to be processed and the demand for support staff increases. Large bureaucratic staffs do not normally foster and facilitate initiative and boldness in subordinate organizations.

The military has already experienced the unintended consequences of automation when it comes to bureaucratic staffwork. Consider how the word processor has streamlined typing and the electronic mailing of documents and messages. As stated before, automated offices are able to effortlessly and with efficiency send documents and E-mail via Local Area Networks and, through MILNET, between organizations. These tools obviously save time through simplifying distribution, increasing information transfers between organizations, thereby decreasing the number of secretaries and military clerks. However, the demand for more usable information by senior decision makers has also increased, requiring staff officers to spend greater portions of their time typing questions and answers. With the increase in productivity, automation has fostered an increase in reporting and information generation. It is true that reports and briefs are now written easily on computers, but organizations tend now to spend more time reworking letters and briefing charts to achieve limited gains in redrafts and minor formatting. The reworking does not necessarily result in better content but only ensures cosmetically perfect documents. The labor saving and time saving office computer has the unintended consequence of creating more work. Computers have increased the military staffs work load and have unintentionally required more staff, not less.

Increased Fragility

General Sullivan, Chief of Staff, Army, states that technology-driven battle space awareness "can provide us with lighter, more effective and more lethal weapon systems to offset our smaller force structure."[16] But, one also must consider that as these force-multiplying systems are affected by attritions of war, their corresponding impact upon the force is an equally multiplying loss. Digital technologies provide a tremendous capability; however, they must be weighed cautiously against their vulnerabilities.

The use of digital technology has consequences that are paradoxical in nature. While it is true that the implementation of more advanced sensor-to-weapon systems will give better situational awareness, help minimize fratricide and better focus combat power, they can also be our Achilles heel. There is the concern whether the new electronic systems will have the resilience to absorb shocks and withstand perturbations associated with the realities of modern war. The more advanced the equipment the more complex it becomes. Generally, the more complex the system, the less reliable, and the harder or longer it takes to repair.

A possible solution to the attrition issue is the use of redundancy in the digital architecture. Redundancy is not likely to protect the force as technology is being looked upon by our senior leaders to help mitigate the smaller force structure. It is a reality that the U. S. military will always be dependent upon finite amounts of specialized electronic equipment that will invariably not be easily replaced nor repaired.

There are other compounding effects of attrition, such as the digital technologies dependence upon reliable communication architectures. It is the dependence and reliance of electronically synthesized data from widely dispersed systems that makes communications the most vulnerable segment. The greatest threat to the entire digitization architecture is the lack of robust, redundant communications paths. For example, Desert Storm validated the notion that a CONUS based force projection strategy, with its anticipated tempo is dependent upon assured satellite communications for both inter- and intra-theater communications irrespective of use on land or sea. Above the tactical level of war, it is anticipated for the Twenty-first century that medium data rates for the military, as well as civilian satellite systems, will be vulnerable to jamming by today's low cost technology.[17] Since digital communications demands are going to be predictably greater than their capacity, there will be no redundancy nor unused capacity.[18] Even today, it is doctrine not to normally keep communication equipment in reserve. Also, key communications equipment is subject to targeting through old-fashioned direct or indirect fires. An unsophisticated enemy sharp-shooter could easily fire a well placed bullet and take out the feed to a critical satellite or a microwave antenna.[19] This would effectively disrupt the digital architecture with devastating consequences to the command and control process. Communications will be the weakest link for the future's dispersed digitized systems.

There is a trend growing within the U.S. military to use commercially available off-the shelf technology that is generally not designed to Military Standards nor for ease of trouble shooting. An unintentional result of using the less rugged commercial electronic equipment is it will require warfighting organizations to have near-by repairmen and large stocks within theater of specialized electronic repair parts. Otherwise, replenishment could become a problem since strategic mobility will always be at a premium during a crisis.

Repair parts may not always be available. A CONUS based support structure assumes that the tempo at which critical items are used and destroyed will not exceed the rate at which they are being replaced. If tradition holds, then the military will keep electronic equipment beyond normal commercial applications. CONUS depots will have to have on-hand the repair parts since industry does not normally keep the production lines of specialized nor dated electronic designs operating. Leaders will need to take into account and understand the perishability and limited availability of electronic equipment.

Lack of timely repair could contribute to the increased fragility of the force. The advanced equipment of the future, whether it be built to commercial or military standards, will require a greater level of maintenance expertise. The philosophy of using built-in diagnostic and automatic check-out equipment does design complexity away from the operator; however, repair responsibilities dramatically shift to maintenance personnel.[20] Repairing systems in the forward areas will be more difficult and there will be a greater tendency to evacuate equipment for repairs to the depots; thus increasing the time to repair.[21] Based on complexity, it can be anticipated that digital electronic systems are going to have a much longer logistics tail than we have today.

Our force structure will likely have a critical shortage of qualified electronic technicians. Consider the same scenario of equipment damaged by a "low-tech" sharpshooter. What if the shooter decides to wait around to kill the skilled operator or maintenance person conducting the repair?[22] Skilled repairmen will not be replaced easily. Maintenance personnel of the future will require more extensive and sophisticated training than today's technicians, since the systems will be more intricate. As the armed forces continue to compete against educational institutions and private corporations for the limited number of qualified 17-24 year-olds, it is going to be a challenge to keep qualified personnel skilled in electronics. Even the reliance on dedicated and first-rate National Guard and Reserve units is optimistic. Units meant to augment during national emergencies may not be able to keep operator and repair personnel up to warfighting standards. The implication--"there will be time for training after the troops are assembled--is virtually over."[23] It is not quantitatively known whether the rate of replacement of critical specialty skills will be able to keep pace, but one has only to look at where we are, traditionally short of technical specialists today. As during Desert Storm, it will be common for the Services to depend upon civilian technicians in forward deployed areas for maintenance and repair, due to the maintenance complexity of electronic equipment and the lack of qualified military personnel. It is possible that demographics, and sociopolitical educational realities, might just be the limiting factors of the Twenty-first century digitized force. The demand for "high-tech" personnel is going to go up--not down.

Another concern for the future is that we will depend more upon our allies for support. As the United States modernizes its forces with the latest electronic gadgetry, will our allies be willing to invest in the same kinds of equipment, or will we be willing to sell or give it to them' By digitizing everything from Command and Control to weapons systems, they might make it impossible to be electronically interoperable within a theater with our coalition partners.

Challenge of the Future

It is obvious that embracing digital technology offers a great deal to the warfighter. If the equipment envisioned for the future is developed and fully realized, the enhancement to Command and Control will dramatically streamline operations. For the future, technology will permit greater situational awareness that will reduce the commander's uncertainty and anxiety. It will certainly increase the warfighter's lethality through the ability to better focus and synchronize combat power. The unintended consequences of this solution will be a challenge for future military leaders to understand. Leaders need to understand the impact upon the operational art of war that technology brings.

Digital technology allows many advantages, but it is not without its soft intangible price. The possibility that the very strength that digital technology brings, situational awareness, may be the cause for trimming a commander's prerogatives. This trimming undermines the very concept of individual initiative that makes the U. S. military such a formidable force. Another soft cost is that complex equipment may be so intricate as to require skills and resources that are not easily replaced during the conduct of war.

The U.S. military needs to adapt to new technology or it faces the prospect of allowing itself to grow flaccid and obsolete. The issues raised in this paper have no direct solution. Yet, a great deal of the problems highlighted, especially in the area of commander's prerogatives, is dependent upon how the senior leadership grows to use the technology and how they guide the emerging organizational culture to use it. Awareness of the strengths and, more importantly, the weaknesses that technology brings to an organization must be the precursor to its implementation. Caution and reflection might be in order to insure that a particular solution's negative ramifications are well understood and its advantages are absolute. It is safe to predict that joint digitization and all its capabilities that it brings will be beneficial. It would be unfortunate, however, that in the process of fixing today's problems a whole new array of unanticipated consequences arise without a plan to discern nor understand them. Developing a clear plan is the challenge for the future.

Notes

1. Gen John M. Shalikashvili, "Letter of Introduction," *C4I for the Warrior: Global Command & Control System, from Concept to Reality,* (Washington: J6 Joint Staff, The Pentagon, 12 June 1991). The GCCS objective is to provide "total battle space information to the warrior.

2. Ibid., "Proofs of Concepts."

3. Christopher V. Cardine, "Digitization of the Battlefield," Unpublished Research Paper, U.S. Army War College, Carlisle Barracks, PA: 1991, p. 9.

4. Douglas A. MacGregor, "Future Battle: The Merging Levels of War," *Parameters,* Winter 1992-93, p. 41.

5. Ibid., p. 33.

6. U.S. Department of the Army, *Force XXI Operations: A Concept for the Evolution of Full-Dimensional Operations for the Strategic Army of the*

Early Twenty-First Century, Fort Monroe, VA: Army Training and Doctrine Command, 1 August 1994, p. 2-9.

7. U.S. Department of the Army, *Force XXI Operations: A Concept for the Evolution of Full-Dimensional Operations for the Strategic Army of the Early Twenty-First Century*, Fort Monroe, VA: Army Training and Doctrine Command, 1 August 1991. p. 2-8.

8. Martin Van Creveld, *Technology and War: from 2000 BC to the Present*. 2nd ed. (New York: The Free Press, A Division of Macmillian. 1991), p. 247.

9. U.S. Marine Corps, *Warfighting*, FMFM 1, Washington: U.S. Marine Corps, 1989, p. 52.

10. Antulio J. Echevarria and other, "The New Military Revolution: Post-Industrial Change," *Parameters*, Winter 1992-93, p. 77.

11. Frank J. Stech, "Winning CNN Wars," *Parameters*, Autumn 1994, p. 39.

12. Echevarria, p. 77.

13. Telephone conversation with LtCol Basla, The Joint Chiefs of Staff (J6V), Washington, DC, 27 January 1995.

14. Cardine, p. 23.

15. Van Creveld, p. 236.

16. Gordon R. Sullivan, "Moving into the 21st Century,: America's Army and Modernization," *Military Review*, July 1993, p. 6.

17. Medium data rates are considered above 2.4kbs to 1.544Mbs. The Milstar and UFO-follow-on satellite systems have EHF antijam systems that will provide the tactical & strategic commanders with assured communications. However, part of Milstar is designed specifically for the *tactical* commander and is the only satellite that can provide EHF antijam medium data rates above 2.4kbps up through to 1.545Mbs (Tl-rate). Other systems such as DSCS and the UHF systems cannot assure large anti-jam capacities at the anticipated loading. It should be noted that DSCS is capable of high data rates in a benign, nonjamming environment.

18. Telephone Conversation with LTC Primo, The Joint Chiefs of Staff (J6S), Washington, DC, 25 January 1995. According to LTC Primo, demand will continue to exceed capacity; satellite data requirements are growing exponentially. In a recent study data rates for emerging requirements have approached SGbs in comparison to today's satellite capacity potential of 1.5Gbs. It should be noted that potential satellite capacity assumes perfect and optimum link closures using large

strategic dishes (60'). The use of tactical or "disadvantaged" smaller dishes (with 8' or smaller dishes) *significantly* reduces the capacity available to the user.

19. Julie Ryan and et al., *Information Support to Military Operations in the Year 2000 and Beyond: Security Implications,* (Alexandria, VA: Center for Naval Analyses, November 1993), p. 11.

20. Martin Binkin, *Military Technology and Defense Manpower.* (Washington: The Brookings Institution, 1986) p.58.

21. An example of shipping complex electronic equipment back to the depot for repair is the Precise, Lightweight GPS Receiver (PLGR). Units are shipped back to the manufacturer under a military procured warranty program. No repairs are attempted in the field. This method does insure first class repair, but it temporarily removes the receiver out of the hands of the user unless a logistics "float" is available within the unit.

22. Ryan, et al., p.11.

23. Echevarria, p. 76.

Bibliography

Binkin, Martin, *Military Technology and Defense Manpower.* Washington: The Brookings Institution, 1986.

Cardine, Christopher V., "Digitization of the Battlefield," Unpublished Research Paper, U.S. Army War College, Carlisle Barracks, PA 20 May 1994.

Campen, Alan D. *The First Information War.* Fairfax, VA: AFCEA International Press, 1992.

Echevarria, Antulio J. and Shaw, John M., "The New Military Revolution: Post-Industrial Change," *Parameters*, Winter 1992-93, pp. 70-79.

MacGregor, Douglas A., "Future Battle: The Merging Levels of War," *Parameters*, Winter 1992-93, pp. 44-47.

O'Berry, Carl G. *Horizon.* Washington: Air Force Deputy Chief of Staff, Command, Control, Communications, and Computer Plans and Policy Division, The Pentagon.

Ryan, Julie; Federici, Gary; Thorley, Tom; *Information Support to Military Operations in the Year 2000 and Beyond: Security Implications,* Center for Naval Analyses, Alexandria,VA: November 1993.

Shalikashvili, John M. *C4I for the Warrior: Global Command & Control System, from Concept to Reality*, The Joint Chiefs of Staff(J6), Pentagon, Washington: 12 June 1994.

Stech, Frank J., "Winning CNN Wars," *Parameters*, Autumn 1994, pp. 37-56.

Sullivan, Gordon R., "Moving into the 21st Century: America's Army and Modernization," *Military Review*, July 1993, pp. 2-11.

Sullivan, Gordon R. *Army Enterprise Strategy: The Vision.* Washington: U.S. Department of the Army, 20 July 1993.

Sullivan, Gordon R. and Dubik, James M. *War in the Information Age.* Carlisle Barracks, PA: Strategic Studies Institute, U.S. Army War College, 6 June 1994

Telephone conversation, LtCol Basla, USAF, Global Command & Control System Project Officer, The Joint Staff(J6V), Pentagon, Washington, DC, 27 January 1995.

Telephone conversation with LTC Primo, USA, Staff Officer, J6(S), Pentagon, Washington, DC, 25 January 1995.

Tuttle, Jerry O. *Copernicus.* Washington: U.S. Department of Navy, 1993.

U. S. Department of the Army. *Force XXI Operations: A Concept for the Evolution of Full Dimensional Operations for the Strategic Army of the Early Twenty-First Century.* TRADOC Pamphlet 525-5. Ft. Monroe, VA: 1 August 1994.

U.S. Marine Corps. *Warfighting.* Marine FMFM 1. Washington: U.S. Marine Corps, 6 March 1989.

Van Creveld, Martin. *Command in War.* Cambridge, MA: Harvard University Press. 1985.

Van Creveld, Martin. *Technology and War: from 2000 B.C. to the Present,* New York, NY: The Free Press, A Division of Macmillan, Inc. 1991.

○ INFORMATION WARFARE:
Issues and Perspectives

Dr. John H. Miller

One of the most intriguing ideas currently circulating in the defense community is the concept of "Information Warfare." The precise meaning of the term is elusive, in part because it describes a wide range of seemingly unrelated phenomena. As currently used, it can refer to everything from computer viruses to "smart" bombs, and encompass the activities of people as diverse as computer hackers and professional soldiers. The ambiguity is increased by the common tendency to employ the term more or less interchangeably with a variety of others such as "Cyberwar," "Netwar," "Third Wave Warfare!" "Command Warfare" and "Post-industrial Warfare"—not all of which mean the same thing.[1]

Part of the difficulty stems from the fact that information Warfare embraces several related, but distinct sets of ideas which are not always clearly distinguished. For many defense analysts, it refers primarily the military application of computers and other information technologies, and the organizational, operational and doctrinal changes this implies for the U.S. and other military establishments. For other writers, however, Information Warfare is a much broader idea, relating to the emergence of "Information Age" civilization and the development of associated modes of political and social conflict which point toward the gradual erosion of nation-states and their monopoly of organized violence.

The Information Revolution

Viewed from latter perspective, Information Warfare raises basic questions not only about how wars will be waged but who will wage them and for what purposes. Such questions cannot, in turn, be separated from consideration of the larger changes that may flow from the ongoing "Information Revolution." If futurists like the Tofflers and John Naisbitt are right the global diffusion of new and emerging information technologies will have economic, social, cultural and political consequences as profound as any in human history. It is, of course, impossible to be sure about the direction of these changes, but they could involve dramatic shifts in political power and attitudes toward authority.[2]

A comparison with the political effects of the "First Information Revolution" in Early Modern Europe may be instructive. As J.M. Roberts points out, the invention of movable type and the subsequent diffusion of printed books and literacy gave rise to a "transformation of the European consciousness" after 1500.[3] Among its other consequences, this transformation abetted the rise of nation-states by providing their rulers with opportunities to mobilize national loyalties and develop centralized administrations. The long-term political "losers" were subnational and supranational entities such as the Papacy and feudal authorities who proved less efficient at exploiting the new medium.

The present Information Revolution may be having precisely the opposite effect inasmuch as the globalization and personalization of electronic communications system appear to be undermining the authority of nation-states and facilitating a devolution of power to subnational and transnational movements, especially those that tap ethnic, religious or cultural loyalties. Naisbitt, who regards such movements as manifestations of a "new tribalism" speculates that traditional institutions of central government and representative democracy may become increasingly anachronistic even in Western societies as demands there for "self rule" and decentralization gain momentum.[4]

One should not, of course, exaggerate the "empowerment" of the individual vis-a-vis the state. As David Ronfeldt notes, the Information Revolution "may give a state apparatus and its rulers powerful new means of control over their citizenry, with an official ideology determining what information is allowed."[5] This may be less of a concern in the United States and other countries where democracy has deep roots, but it is a distinct possibility elsewhere, particularly in the Third World where it is easier for charismatic leaders to generate a public consensus in favor of tyranny. The Information Revolution may promote either democracy or totalitarianism depending on the socio-political context.

It is likewise premature to conclude that the presumed decline of nationstates will inhibit conflict and war. Indeed, the Information Revolution may actually stimulate conflict by accentuating economic, cultural and political differences among peoples while at the same time binding together groups previously separated by geographic or national barriers. Benjamin Barber, for example, sees a threatened "Lebanonization" of many nation-states, in which culture is pitted against culture, people against people, tribe against tribe -- a Jihad in the name of a hundred narrowly conceived faiths against every kind of interdependence, every kind of artificial social cooperation and civic mutuality."[6]

While forecasts of a "coming anarchy" or "clash of civilizations" may be overdrawn,[7] war in the Information Age could well spill outside of the Clausewitzian framework where it functions as a "rational' instrument of state policy. John Keegan reminds us that, historically, different cultures have shaped war into bizarre and self-destructive forms whose warrior practitioners, unlike modern soldiers, often looked upon combat as a means of self-expression, recreation or religious sanctification.[8] In some parts of the world, the weakening of state authority and the eruption of violent sociopolitical conflicts are enabling such warriors to make a comeback. "as brutal as ever and distinctly better armed."[9]

In his chilling vision of war in a post-Clausewitzian world Martin van Creveld takes this line of reasoning a step further, speculating that modern high-tech armies are like to become obsolete. In his view, the relevance of these armies increasingly will be called into question because they will be unable to decisively defeat guerrillas and terrorists who operate beneath their "sophistication threshold" in low-intensity conflicts. Future wars, he suggests, will not be clean and short but "protracted, bloody and horrible" -- an affair of "listening devices and of car bombs, of men killing each other at close quarters, and of women using their purses to carry explosives and the drugs to pay for them."[10]

Van Creveld's work carries another interesting and, for professional soldiers, more disturbing implication: In order to wage low-intensity conflicts with any hope of success, conventional armies may have to adopt the organizational methods, and perhaps even the mentality of their opponents. Noting that war represents the most imitative activity known to man." van Creveld predicts that pervasive low-intensity conflict "will cause regular forces to degenerate into police forces or, in case the struggle lasts for very long, mere armed gangs."[11] Distinctions will thus erode between military and police forces, and ultimately between soldiers and the terrorists and criminals whom they are responsible for combating.

It might be objected that van Creveld and others who see non-Clausewitzian war as the wave of the future underestimate the political dimension of the violence occurring in places like Bosnia. The systematic massacre or brutalization of non-combatants may affront "civilized" sensibilities and conceptions of warfare. If employed in the pursuit of political objectives such as the creation or defense nation-states, however, even "ethnic cleansing" can be seen as a continuation of politics.

The problem, of course, is where to draw the line. Clausewitz himself recognized that war is at bottom an affair of "primordial violence" whose inherent tendency is to run to extremes of emotion and cruelty.

One can also question the strategic significance of many low-intensity conflicts, especially in the post-Cold War context. As appalling as their human costs may be, few of the small wars currently underway in the Third World and former Second World engage the vital interests of major powers or seem likely to bring about immediate changes in the international balance of power. The United States may choose to involve itself in military efforts to try to resolve or contain these struggles. With the disappearance of the Soviet Union as a global competitor however the geopolitical rationale for doing so has lost much of its former force, and the American interests at stake are often highly ambiguous.

"From a purely American point of view," as Eliot Cohen observes, "the world is, and for some length of time promises to be, a more secure place than it was during the cold war."[12] American statesmen now have, he notes "no dragons to slay, or even to tame; partners may be competitors, but no state poses a direct challenge to our security or that of our allies." Even with the current downsizing of the U.S. military establishment, its Cold War "capital stock" of sophisticated military platforms, intelligence-gathering systems, military-technical infrastructure, and logistical assets, dwarfs that of any potential competitor and gives the United States unmatched global power-projection capabilities.

Although the external threats currently facing the United States are more diffuse and ill-defined than during the Cold War era, there is no lack of potential dangers and challenges. Clearly the most menacing contemporary development is the proliferation of weapons of mass destruction. Only slightly less worrisome, however, is the diffusion of long-range missiles, advanced aircraft, and other high-tech weaponry Both trends are likely to enhance the military capabilities of state and non-state actors, thus increasing the escalatory potential of ethno-sectarian and inter-state conflicts. These trends may also make it possible for even relatively weak powers to strike directly at the United States.

The Military-Technical Revolution

The ongoing Information Revolution is having equally important, if less obvious effects on the post-Cold War security environment. There is, for example, broad agreement in the defense community that the 1991 Gulf War marked the early stages of a "Military-Technical Revolution" (MTR) which promises to transform the character of war as radically as did the advent of nuclear weapons fifty years age. Driving this revolution are advances in surveillance, communications, and information-processing technologies, which create the possibility of imbuing the "information loop of warfare" with unprecedented accuracy and speed, thereby achieving "information dominance" over less capable adversaries.[13]

Information dominance, combined with cruise missiles and other precision weaponry, is expected to confer the ability to overwhelm virtually any opponent quickly, decisively and relatively bloodlessly. The key problem in this mode of warfare will be protecting one's own "centers of gravity" or vulnerabilities while striking those of an enemy. As was foreshadowed in the Coalition air campaign in the Gulf War, the solution to the latter problem is likely to involve "an improved ability to understand target systems and their relationship to operational and strategic objectives,"[14] since knowing which targets to strike will be critical to the effective employment of large numbers of precision weapons.

Full exploitation of information technologies will, it is argued, necessitate major changes in "the ways militaries think about, organize themselves for, and wage combat."[15] Thus, reliance on sensors, computers and smart weapons points toward

150

a reduced human role in decision cycles. Similarly, long-range strike capabilities will lessen the need for close-in combat and increase the importance of integrated air-land-sea-space operations. The extraordinary lethality of precision weapons and their ability to strike virtually anywhere will likewise require dispersed, independent combat units, and the replacement of present types of aircraft armored vehicles and surface ships by smaller and more stealthy platforms.[16]

Beginning with then Undersecretary Perry's announcement in the aftermath of Desert Storm that the U.S. had achieved a revolutionary advance in military capability,"[17] many U.S. military leaders have enthusiastically embraced the MTR idea. The main doctrinal construct for implementing it at operational level is the strategy of "Command and Control Warfare," which calls for "the integrated use of operations security, military deception, psychological operations, electronic warfare, and physical destruction, mutually supported by intelligence, to deny information to influence, degrade or destroy adversary C2 capabilities, while protecting friendly C2 capabilities."[18]

Enthusiasm within the U.S. military for the MTR is tempered by concerns about over-reliance on technology and the possible loss of traditional combat skills.[19] But skepticism about the MTR in professional military circles has deeper roots. As C. Kenneth Allard points out, the integration and information sharing required by emerging command and control technologies threatens to upset the "delicate balance" of service autonomy and, ultimately, the very notion of military hierarchy itself.[20] Despite universal acceptance of the twin imperatives of "jointness and interoperability," service cultures and perspectives are therefore likely to act as a brake on institutional change.

For many MTR proponents, the U.S. military has barely begun the "intellectual revolution" necessary to realize the quantum leap in combat effectiveness potentially offered by information technologies. Dan Goure, for example, compares the present level of institutional adaptation to the U.S. Navy's initial efforts to develop carrier aviation in the 1920's. Absent a constituency for more radical doctrinal and organizational reforms, he predicts that U.S. military planners will "fall prey to the old tendency to look to technology as the solution to strategic and operational problems" and settle for the "fleeting advantage" conferred by "mere technological superiority."[21]

According to Goure and others, this piecemeal approach may create opportunities for more innovative competitors to vault ahead of the United States. Based on his study of previous military revolutions, Andrew Krepinevich warns that it is a "dangerous delusion" to assume that the United States will be able to control the direction of the current revolution.[22] "Even when countries will not be able to compete in the full spectrum of military capabilities," he argues, "some of them, by specializing, will become formidable niche competitors." Moreover, they may find it easy to do so since the Information Revolution is lowering the cost of information technologies and increasing their availability.

In a similar vein Paul Bracken urges U.S. military planners to devote more attention to what he sees as the likely emergence 15 or 20 years hence of "Eurasian peer competitors." According to bracken, the rapid economic growth of countries like China will give them the means to field modern military forces based on advanced technologies and innovative operational concepts. Unless the United States begins now to think seriously about how to counter such challenges, it risks becoming "locked into" current force structures, concepts technologies and doctrines which—although sufficient to deal with near-term contingencies—may be inadequate to sustain the long-term U.S. competitive advantage.[23]

A basic problem with these scenarios is that they rest on assumptions about the future shape of the international order, the nature of war, and the direction of technological change, none of which can be taken for granted. As Goure notes, military planners during the Cold War and interwar periods faced the task of bringing "known pieces of potential capability together in a fashion suited to an expected form of warfare against anticipated adversaries.[24] Under current conditions of rapid technological change and the emergence of new subnational and transnational players, however, such certainties are denied planners who cannot be sure about the form or participants that will characterize future wars.

Since the MTR is seen as a long-term process and presupposes threats which have not yet materialized, its relevance to current defense needs is open to question. The basic rationale for reshaping the U.S. military now comes down to providing a hedge against the possible rise of future high-tech challengers. In addition, pushing ahead with the MTR is espoused as a way of increasing the effectiveness of U.S. forces in dealing with near-term, regional contingencies and conducting military operations short of war.

Domestic, political, and budgetary constraints, which require the U.S. military to "do more the less" and minimize casualties on all sides, are also cited to justify this course.[25]

The adequacy of current U S. military plans and programs to accomplish these objectives is a subject of controversy. The 1993 "Bottom-Up Review." DOD's blueprint for meeting the challenges of the post-Cold War security environment. emphasizes the need to maintain military capabilities sufficient to defeat regional aggressors in two "nearly simultaneous" conflicts. The areas of particular concern are the Korean Peninsula and the Persian Gulf, and the most likely aggressors are deemed to be North Korea, Iran or a resurgent Iraq. The kind of attack foreseen (an armor-heavy, combined arms offensive") and the forces considered necessary to defeat it are modeled on the 1990-91 Gulf conflict.[26]

Many critics contend that the Bottom-Up Review falls into the familiar trap of "preparing to fight the last war" instead of the next one.[27] They argue, for example, that regional aggressors are unlikely to repeat Saddam Hussein's blunders and will instead employ a range of strategies to deter U.S. military intervention, including "ambiguous aggression," guerrilla warfare, and threats to use weapons of mass destruction. It is also suggested that the kind of forces appropriate to counter threats from North Korea or Iraq are not suitable for meeting future challenges from more formidable peer competitors who may succeed in combining information technologies with innovative military doctrines.

Another focus of criticism is the force mix recommended by the Bottom-Up Review to maintain U.S. power-projection capabilities. Edward Luttwak is typical of those who see this mix as reflecting the success of the services in preserving their Cold War force structures rather than shifting to "low-casualty or noncasualty forms of aerial and robotic military strength" more suited to the post-Cold War security environment.[28] Luttwak and others are also concerned that the costs of supporting carrier battlegroups and heavy army divisions will consume an undue share of DOD's shrinking budget, allowing fewer resources for R&D and modernization programs necessary to sustain the U.S. comparative military advantage.

Some critics are also skeptical that it will be possible to reconcile the Bottom-Up Review's regional conflict strategy with the growing demands imposed on the U.S. military by peacekeeping, peace enforcement and other "operations other than war."[29] Since these operations tend to be manpower-intensive, expensive, and protracted, they drain resources needed to assure the readiness and modernization of forces which might be required to respond to major regional contingencies. Moreover, irregular operations, by their nature, offer fewer opportunities than conventional warfare to employ U.S. surveillance, command and control, and precision strike capabilities to achieve decisive military results.

The prescriptions for these problems offered by MTR advocates range from more rapid structural reform of the U.S. military to recasting American defense strategy. Krepinovich, for example, urges that higher priority be accorded to "preserving the long-term military potential of U.S. forces, as opposed to near-term capabilities."[30] This implies accepting a smaller force structure and devoting more resources to modernization. Such an approach also entails greater use of multinational coalitions to deal with regional contingencies, increased reliance on stand-off precision strikes to punish or deter troublemakers and more selective engagement of U.S. forces in peacekeeping and humanitarian ventures.

The emphasis put by Krepinovich and others on preparing for future high-tech conventional wars has not gone unchallenged. A.J. Bacevichi, for example, charges that the MTR idea is at bottom an exercise in "wishful thinking" designed to relieve U.S. military leaders of the burden of coming to grips with the realities of nuclear and unconventional war for which "they never devised an adequate response."[31] In a similar vein, Daniel Bolger, himself an infantryman, cautions that the "sanguinary shade" of Vietnam may rise again unless the US. Army "forsakes the seductive urge to keep refighting World War II" and devotes more attention to basic infantry skills needed to combat irregular opponents.[32]

MTR theorists do not entirely ignore unconventional warfare but it is not central to their concerns. A recent study of the subject concludes that advanced surveillance and precision-strike systems can at present make only "limited contributions" to irregular operations, owing to the complexity of the human and natural environments in which such operations often must be conducted. While holding out the prospect that emerging sensor and non-lethal weapons technologies may change this situation in the near future, its authors recommend that priority should be given to "traditional combined-arms operations" which "pose the most serious risk to U.S. interests" over the next 10-15 years.[33]

Several factors may, however, increase the relative importance of irregular warfare in the post-Cold War setting. As noted above U.S. dominance of the conventional battlefield is likely to encourage aggressive regional powers to eschew blatant adventurism in favor of indirect strategies, including sponsorship of low-intensify conflicts. Furthermore, cheap commercial information technologies such as satellite-based personal communications devices, encrypted fax machine, and global positioning systems will provide "force multipliers" for irregular forces. The availability of sophisticated weapons from the former Soviet bloc and other suppliers will also enhance the their capabilities.

The growing dependence of modern societies on computer-controlled information infrastructures is opening up new areas for unconventional warfare. Recent intrusions into the U.S. civilian and military computer networks by hackers and criminals only hint at far more serious disruption that could be wrought by systematic electronic sabotage of national telecommunications systems power grids and financial institutions. As information security specialist Winn Schwartau suggests, such sabotage is a 'low risk/high reward endeavor" inasmuch as it can be conducted by "remote control" and offers the prospect of inflicting "indiscriminate damage on millions of people with a single keystroke."[34]

The advent of satellite-based global television broadcasting has created yet another arena for unconventional warfare. As was demonstrated in the Gulf War and the 1993 skirmishing in Somalia between UN forces and the clan warriors of General Aidid, live television coverage provides participants in armed conflicts with unprecedented opportunities to conduct military deception and shape the way distant audiences perceive events on the battlefield. Such "CNN Wars" are likely to become more common and to have disproportionately large political repercussions, especially in societies like the United Slates where policymaking is sometimes driven transitory public reactions to media images.

The foregoing trends point to changes not only in the environment in which wars are fought but in the nature of warfare itself. Information technologies are enlarging the venue of conflict from the traditional "battlefield" to computer networks, electronic databases, television screens, and other unconventional settings. They are also expanding opportunities for waging war through sabotage, terrorism, propaganda and similar clandestine means. One possible consequence is that it may be increasingly difficult to draw a clear line between war and other forms of conflict. Another is that distinctions between military and civilian practitioners of war may become progressively blurred.

Information Warfare

What are the implications of these trends for the MTR concept and the future of professional military establishments? Following van Creveld, one might conclude that, since modern high-tech armies are ill-suited to waging unconventional wars, they will either wither away or be transformed into paramilitary security forces. If, on the other hand, one accepts the premise that technology-based information superiority confers the ability to dominate any opponent, a military with this advantage should be able to prevail in unconventional wars. What is needed, however, is a conceptual framework that identifies strategies and requirements for achieving such dominance across the full spectrum of conflict.

DOD has begun to move in this direction by expanding its doctrine of Command and Control Warfare into the broader concept of Information Warfare (IW). According to a provisional DOD definition, IW consists of "actions taken to achieve information superiority in support of national military strategy by affecting adversary information and information systems while leveraging and protecting our own information and information systems."[35] This formulation is intended to encompass military and non-military actions as well as defensive and offensive aspects. It also covers all levels of war from the tactical to strategic, and applies to troth peacetime and wartime conditions.

According to writers like Donald Ryan, IW is based on the assumption that information technologies have developed to the point where they can now be employed as a weapon in their own right rather than as a "handmaiden" of armed combat.[36] It thus becomes possible to envision wars being "won" or "lost" without, a shot being fired. In contrast to Command and Control Warfare, which aims at victory on the battlefield, IW seeks to avoid the need to resort to lethal force by putting "enemies in positions where their information resources are useless or, worse, unreliable." In this respect, IW aspires to realizing Sun Tzu's famous dictum that "to subdue the enemy without fighting is the acme of skill."

Employed offensively, IW emphasizes the manipulation of electronic information systems to influence an adversary's perceptions and behavior. This might, for example, involve disabling military and civilian telecommunication systems through computer viruses or electromagnetic pulse devices. Infiltration is, however, the "maneuver of choice" since an enemy, unaware that his information sources have been compromised, will continue to trust them, creating opportunities for deception.[37] Offensive IW also emphasizes the use of direct broadcast satellites, the commercial media, and "visual stimulus and illusion" technologies such as holography to conduct propaganda and subversion.

Defensively, IW requires an ability to detect and thwart attempts to tamper with one's own sources of information. In the military sphere, this entails assuring the integrity of command and control, communications, and intelligence systems. Critical elements of the civilian infrastructure such as power grids, financial networks, and telecommunications systems must also be protected. In addition, IW posits an ability to counter enemy propaganda and disinformation. According to the Tofflers, this can best be accomplished in an Information Age context by "precision-targeting" audience segments, disguising propaganda as news and entertainment, and employing computer-generated special effects.[38]

IW theorists like John Arguilla set great store by the techniques of systems analysis to identify enemy and friendly centers of gravity.[39] This approach, which is modeled on the air campaign plan against Iraq, visualizes a state or society as a set of systems and subsystems linked by numerous "nodes." Given sufficient knowledge of such systems, one can isolate the 'critical nodes ' which when destroyed or disabled will cause a systemic collapse. In most cases, the political leadership will be the critical node of the entire state structure, but if it cannot he reached with hard or "soft kill" weapons, belief systems, economic infrastructures, and military forces may have to be attacked.

Applied in an IW context, the systems approach presupposes not only comprehensive knowledge about an adversary, but an ability to continually update that knowledge so that one is in a position to react more swiftly to a changing situation. Such "situational awareness" depends, in turn, on the use of artificial intelligence and other advanced data processing technologies to "fuse" large quantities of human and technical intelligence into an accurate, real-time picture of events. While most IW theorists acknowledge that anything resembling omniscience is unattainable, they do claim that these technologies confer a potentially decisive advantage over opponents lacking comparable capabilities.[40]

The United States faces both opportunities and challenges in waging Information Warfare. On the one hand, it enjoys a wide lead over potential competitors in adapting information technologies to military purposes, and its private sector is at the cutting edge of the global Information Revolution. At the same time, however, the American information infrastructure, on which U.S. defense communications depend, is highly vulnerable to infiltration and sabotage. A recent Joint Security Commission Report describes this vulnerability as "the major security challenge of this decade and possibly the next century" and warns that there is "insufficient awareness of the grave risks we face in this area."[41]

"Hardening" the U.S. national information infrastructure, however, will not be easy. From a purely technical standpoint, it would be a "lengthy and extremely costly" undertaking which might not ensure protection for all types of sensitive date. Moreover, neither DOD nor any other Federal agency currently has the legal authority to enforce information security standards in the private sector, and the imposition of such standards probably would be controversial, especially in the absence of a clear external threat to national security. Despite public concern over privacy and computer crime, reaping the benefits of the "information superhighway" is widely believed to require openness and accessibility.[42]

It will also not be easy for the U.S. to wage Information Warfare at the political level. It is unclear, for example, whether DOD or some other Federal entity would have primary responsibility. Since IW involves issues which affect numerous government agencies and cut across the public and private sectors, a case could be made for the establishment of a separate agency, or coordinating body. However, such organizational problems pale in significance compared to the controversy likely to be generated by the official adoption of a strategy which sanctions, or appears to sanction, systematic "perception management" activities aimed at the American people in the name of national security.

Whatever the means employed, it is questionable whether American public opinion can be induced to support risky foreign policy ventures absent a clear threat to U.S. interests. As Richard Haas points out, building consensus around such ventures is extremely difficult in a political environment characterized by a focus on domestic priorities, an assertive Congress, and heightened media scrutiny.[43] The unraveling of U.S. policy in Somalia after the October 1993 firefight with

Aidid's followers illustrates the problems that can arise in this kind environment, especially when the executive branch neglects to build a reservoir of public support by explaining the costs and risks of its policies.

Information Warfare also poses potential foreign policy and international legal complications. Eliot Cohen, for example, worries that "a military fighting the shadowy battles of "information warfare" may find itself engaging the country in foreign policy tangles of a particularly messy kind."[44] IW operations undertaken in peacetime by the United States for purposes of deterrence or compellence, such as electronically sabotaging the power grid of a rogue state, might well be considered "terrorism," exposing the U.S. to international criticism and possible retaliation in kind. Operations of this type could also stiffen the resolve of the leaders and populace of a hostile state rather than coerce them.

The possibility should be considered that Information Warfare may be more effective against some kinds of adversaries than others. In general, the "critical nodes" of nation-states are easier to identify than those of nonstate entities, especially ones organized as networks and camouflaged within civilian populations. Given "precision intelligence," the vulnerabilities of any group can, in theory, be targeted, and the Information Revolution is enhancing the prospects of acquiring such intelligence through improved surveillance and data processing techniques. As is demonstrated by the growth of computer crime, however, the Information Revolution is also enlarging opportunities for concealment.

According to Robert Steele, the U.S. intelligence community as presently constituted is "virtually worthless" in assuring information dominance over terrorists, criminals, guerrillas and other non-state actors who are able to operate beyond the reach of its technical collection capabilities.[45] To, deal with the threats posed by such groups, Steele argues that radically new approaches to intelligence gathering and analysis are needed, including greater reliance on human as opposed to technical collection methods: closer cooperation among military, law enforcement and private-sector organizations; and a concerted effort to tap "open source" data available from the media, academia and business.

The inconclusive U.S. "war" against foreign drug traffickers suggests, however, that better intelligence by itself may not be enough to "defeat" some non-state actors. Drug trafficking is rooted in intractable political, social and economic conditions—including endemic poverty and corruption in producer countries and the demand for illicit drugs in the United States—which can be addressed only at the margins through interdiction. Thus, while improved intelligence and surveillance methods may enable military and law enforcement authorities to break up some trafficker organizations and raise the cost of doing business for others, trafficking itself is unlikely to be stamped out by such methods.[46]

It might be objected that "raising the cost of doing business" is a viable objective in many conflict situations, and Information Warfare offers the U.S. a means of achieving it at a lower human and material cost than reliance on military force alone. It is not difficult. for example, to envision U.S. forces successfully using IW techniques in "peace operations" to calm civil disturbances or induce the participants to settle their differences by non-violence means. Even when armed conflicts cannot be averted or resolved, moreover, these techniques might be helpful in weakening the aggressive capabilities of a target group by sowing distrust among its members or undermining its base of popular support.

By the same token, however, the United States itself presents an inviting target for Information Warfare since its "centers of gravity" are highly visible and difficult to defend. As noted above, these include the vulnerability of the civilian information infrastructure to electronic attack and the tenuous nature of domestic support for foreign military interventions. The sensitivity of American public opinion to military casualties, collateral damage, and the taking of hostages, is another obvious pressure point which foreign adversaries—aided by media coverage of such events—have sometimes manipulated to coerce U.S. policymakers and will no doubt continue to try to exploit.

In sum, the United States faces significant constraints and handicaps in waging Information Warfare despite its undoubted military and technological process. Most of those who advocate the adoption of an IW strategy by the U.S. deal with this subject within a military frame of reference, ignoring or glossing over its broader political, legal, foreign policy, ethical and cultural implications. Critical questions therefore tend not to be addressed. Perhaps the most fundamental of these is whether the American people are willing, in the absence of a clearly defined external threat, to accept the increased government intervention and regulation that would be required to implement such a strategy.

If waging Information Warfare is problematic for the United States, it also poses difficulties for other states, including totalitarian regimes. Such regimes are less inhibited by legal, political and moral constraints than democratic governments. But even a garrison state like North Korea cannot completely protect its infrastructure against electronic manipulation and disruption. Moreover, the tight bureaucratic controls needed to sustain totalitarian system exact a heavy price in economic efficiency, national competitiveness and, ultimately, living standards. The 1989-1991 collapse of most communist states indicates that discontents bred by these systems contains the seeds of their own demise.

As John Patrick suggests, Information Warfare may be best suited to non-state entities like guerrilla movements, drug cartels, and terrorist networks.[47] Such groups have several advantages over formal military organizations, including a decentralized organization and an ability to blend in with their surroundings. They are also adept at manipulating ideas and information to defeat opponents politically. Furthermore, the Information Revolution is enhancing their capabilities by making them more competitive in terms of firepower, increasing their ability to conduct "pyschological operations" through the media, and enabling them to inflict "electronic Pearl Harbors" on vulnerable nation-states.

Conclusions

One is thus led back to the issue posed at the outset of this essay—whether the advent of the Information Age marks the beginning of the end of the monopoly of war by nation-states and their professional military establishments. This issue is largely ignored in discussions of the Military Technical Revolution, which assume that information technologies are merely the latest in a series of innovations which have revolutionized the character of armed conflict among nation-states. The possibility raised by van Creveld that larger changes may be underway in the nature of war itself with which nation-state armies are ill-prepared to cope is thrust into the background or even dismissed as "elegant irrelevance."[48]

Information Warfare theorists, on the other hand, do address the broader societal impact of the Information Revolution and the challenges this poses for conventional militaries. As suggested above, however, they tend to frame these challenges in narrowly technological and military terms, with the result that they often fail to take into account the political, cultural and legal dimensions of the subject. As long as Information Warfare continues to be viewed chiefly as a problem of military modernization, rather than as part of a larger global transformation affecting every facet of society, it will be difficult to transcend the limitations of this approach and to formulate more persuasive recommendations for coping with it.

A precondition for addressing the challenge of Information Warfare is recognizing that there is a problem. This requires moving beyond the horizons of the MTR concept in which dealing with today's low-intensity conflicts is seen as a "small but irksome" task "compared to potentially more critical but totally unknown tasks that may face the nation two decades from now."[49] A second requirement is acknowledgment that the ability of the U.S. government and military to "solve" the problem may be limited. Technology is unlikely to provide all the answers and the American political environment is not conducive to peacetime mobilization for "total war" against still largely hypothetical threats.

NOTES

1. See Antulio J. Achevarria and John M. Shaw, "The New Military Revolution: Post-Industrial Change," *Parameters.* 22, (Winter 1992-93). pp 70-79; John Arquilla and David Ronfeldt, *Cyberwar in Coming.* Santa Monica: Rand, 1992; Alvin and Heidi Toffler, *War and Anti-War: Survival at the Dawn of the 21st Century.* Boston: Little Brown and Co., 1994, and John Arquilla, "The Strategic Implications of Information Dominance," *Strategic Review.* Summer 1994, pp. 24-30.

2. Alvin and Heidi Toffler, pp. 242 ff; John Naisbitt, Global Paradox: *The Bigger the World Economy, the More Powerful its Smallest Players.* New York: William Morrow and Co., 1994, pp. 271 ff. Also, Steven Bankes and Carl Builder, Seizing the Moment Harnessing the Information Technologies, Santa Monica: 1992, pp. 13-18.

3. J. M. Roberts, *The Penguin History of the World.* Harmondsworth: Penguin Books, 1980, pp. 504-5. Also, Elizabeth L, Einstein, "Some Conjectures About the Impact of Printing on Western Society and Thought: A Preliminary Report," *Journal of Modern History,* (March 1968), esp. pp. 19-21.

4. Naisbitt. pp. 42-46. Nation-states are not, of course, disappearing. Naisbitt notes (p.30) that, paradoxically, "as the importance of the nationstate recedes, more of them are being created."

5. David Ronfelt, *Cyberocracy, Cyberspace, and Cyberology: Political Effects of the Information Revolution,* Santa Monica: Rand, 1991, p 63.

6. Benjamin Barber, "Jihad Vs. McWorld," *The Atlantic Monthly* (March 1992), pp. 53-63.

7. Samuel P. Huntington, The Clash of Civilizations?," *Foreign Affairs.* (Summer 1993), pp. 22-49; Robert D. Kaplan, "The Coming Anarchy!" *Atlantic,* (February 1994), pp. 44-76.

8. John Keegan, *A History of Warfare*, New York: Alfred A. Knopf, 1993, esp. Chapter 2.

9. Ralph Peters. "The New Warrior Class." *Parameters*, 24 (Summer 1994). p. 16.

10. Martin van Creveld, *The Transformation of War*, New York: Free Press, 1991, p.212.

11. *Ibid.*, p. 207.

12. Eliot Cohen, "What To Do About National Defense," *Commentary*, (November 1994), p.23.

13. Michael J. Mazarr, *The Military Technical Revolution*, Washington, D.C.: CSIS, March 1993, pp. 23-26.

14. Andrew F. Krepinevich, "Cavalry to Computer: The Pattern of Military Revolutions," *The National Interest* . (Fall 1994), p. 41.

15. Dan Goure, "Is There a Military-Technical Revolution in America's Future?." *The Washington Quarterly*. (Autumn 1993), p. 179.

16. Mazaar, pp. 34-35.

17. William J. Perry, "Desert Storm and Deterrence," *Foreign Affairs*, (Fall 1991), p.66.

18. Joint Chiefs of Staff Memorandum of Policy (MOP) 30.

19. "Battle Plans for a New Century," *The Washington* Post, February 21, 1995, A1.

20. C. Kenneth Allard, *Command, Control and the Common Defense*, New Haven: Yale University Press, 1990, pp. 250, 263-64.

21. Goure. p 182.

22. Krepinevich, p. 42.

23. Paul Bracken, "The Military After Next," *The Washington Quarterly*, (Autumn 1993), pp. 168-69.

24. Goure, p. 183.

25. Mazaar, pp. 10 ff.

26. Quoted in Andrew F. Krepinevich, "The Clinton Defense Program: Assessing the Bottom-Up Review," *Strategic Review*. (Spring 1994), pp. 18-19.

27. Ibid., pp. 19-20. Also, Cohen, p. 24, and Charles T. Allen, "Extended Conventional Deterrence: In from the Cold and Out of the Nuclear Fire?," *The Washington Quarterly*. (Summer 1994), pp. 225-28.

28. Edward N. Luttwak, "Washington's Biggest Scandal." *Commentary*. (May 1994), p. 32.

29. For example, Cohen, p. 25.

30. Krepinevich, "The Clinton Defense Program," p. 24.

31.) A.J. Bacevich, "Preserving the Well-Bred Horse." *The National Interest* (Fall 1994). pp. 48-49.

32. Daniel P. Bolger, "The Ghosts of Omdurman," *Parameters*. (Autumn 1991), p.39.

33. Mazaar, pp. 54-55.

34. Winn Schwartau. *Information Warfare, Chaos on the Electronic Superhighway*. New York: Thunder's Mouth Press, 1994, pp. 21-22.

35. Quoted in *Defense Science Board Summer Study on Information Architecture for the Battlefield*, October 1994, p. B-6 (hereinafter cited as DSB Final Report).

36. Donald E. Ryan, "Implications of Information-Based Warfare," *Joint Forces* Quarterly (Autumn-Winter 1994-95). p. 114.

37. *Ibid.*, p. 116.

38. Alvin and Heidi Toffler, pp. 165-75.

39. John Arquilla pp. 27-29.

40. *Ibid.*, p. 27. Also, Dan Kuehl, "Target Sets for Strategic Information Warfare in an Era of Comprehensive Situational Awareness," Unpublished Paper, January 24, 1995, esp. pp. 9-10.

41. Quoted in DSB Final Report, p. B-5.

42. *Ibid.*, pp. 34-35.

43. Richard N. Haass, *Intervention: The Use of American Military Force in the Post-Cold War World*. Washington, D.C.: Carnegie Endowment, 1994, p. 7.

44. Cohen. p. 31. Also, Keuhl, pp. 7-8.

45. Robert D. Steele, "The Military Perspective on Information Warfare: Apocalypse Now," Keynote Speech to the Second International Conference on Information Warfare, Montreal, Canada, January 19, 1995, p. 13.

46. Peter R. Andreas, Eva C. Bertram, Morris F. Blackman and K.E. Sharpe, "Dead-End Drug Wars," *Foreign Policy*, (Winter 1991-92), p. 112.

47. John J. Patrick, "Reflections on the Revolution in Military Affairs," Unpublished Paper, October 17, 1994, pp. 28 ff.

48. Kenneth F. McKenzie, Jr. "Elegant Irrelevance: Fourth Generation Warfare," *Parameters*. (Autumn 1993), esp. pp. 58-59.

49. Martin Libicki and James Hazlett, "The Revolution in Military Affairs," *INSS Strategic Forum* (November 1994), p.1.

www.ingramcontent.com/pod-product-compliance
Lightning Source LLC
Chambersburg PA
CBHW070800100426
42742CB00012B/2204